San Luis Rey on Display

Geoscience in Northern San Diego County, California

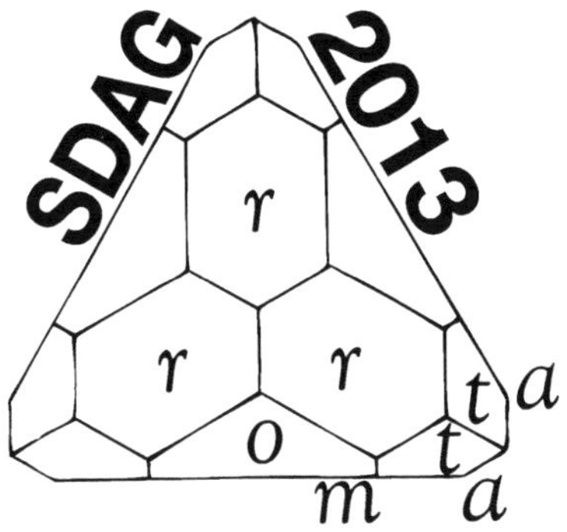

The crystallographic diagram above shows the characteristic rounded triangular cross section of tourmaline, formed by the prominent trigonal prism and subordinate hexagonal prism. Prism faces *r, o, t, a* and *m* round into each other, creating a spherical triangle in cross section. For more information, see Hurlbut and Klein, *Manual of Mineralogy*, page 473.

San Luis Rey on Display

GEOSCIENCE IN NORTHERN SAN DIEGO COUNTY, CALIFORNIA

Published in conjunction with the
2013 Field Trip of the
San Diego Association of Geologists
November 9–10, 2013

Editor: Brian J. Olson
Project Manager: David M. Bloom

SAN DIEGO ASSOCIATION OF GEOLOGISTS

Distributed by Sunbelt Publications
www.sunbeltbooks.com

San Luis Rey on Display:
Geoscience in Northern San Diego County, California

First edition 2013, third printing 2016.

Edited by Brian J. Olson
Cover and Book Design by Michael Schrauzer
Project Manager: David M. Bloom
Printed in United States of America

San Diego Geological Society
3130 North Evergreen Street
San Diego, California 92110
www.sandiegogeologists.org

or the distributor:

Sunbelt Publications
1250 Fayette Street
El Cajon, California 92030
(619) 258-4911, fax: (619) 258-4916
www.sunbeltpublications.com info@sunbeltpub.com

19 18 17 16 6 5 4 3

Library of Congress Cataloging-in-Publication Data
San Luis Rey on display : geoscience in northern San Diego County, California / editor, Brian J. Olson ; project manager, David M. Bloom.
— First edition.
pages cm
"Published in conjunction with the 2013 Field Trip of the San Diego Association of Geologists November 9–10, 2013."
Includes bibliographical references.
ISBN 978-0-916251-36-9 (alk. paper)
1. Geology--California--San Diego County--Guidebooks. 2. San Diego County (Calif.)--Guidebooks. I. Olson, Brian J., editor of compilation.
QE90.S116S26 2013
557.94'98--dc23
2013038879

San Diego Geological Society, Inc., gratefully acknowledges the generous contributions of our sponsors to the publication of this book.
Front Cover: upper right, Joe Budd Photo – The Arkenstone; upper left, David M. Bloom; bottom, San Diego Natural History Museum.
Inside Front Cover: Woodrow L. Higdon, Geo-Tech Imagery International.
Inside Back Cover: Carolyn Glockhoff
Back Cover (clockwise from bottom left): A. Varrett, True, et al. (1974) , San Diego Natural History Museum, Joe Budd Photo – The Arkenstone.

Table of Contents

III. PAPERS

IV. Appendices

Preface

San Luis Rey on Display: Geoscience in Northern San Diego County is a guidebook and collection of papers prepared to accompany San Diego Association of Geologists' (SDAG) 2013 annual field trip. Amongst southern California geologists and San Diego residents alike, San Luis Rey River valley is well known for scenic Highway 76 that winds through its center.

This volume is organized into three parts. The first is a road log for a field trip designed to give participants some intriguing views of the geology, paleontology, and archeology in the valley. The SDAG field trip will take a weekend to see it all. We will begin with a roughly west-to-east trip through what is the *heart* of the elongate drainage basin, extending from California Highway 76-Interstate 15 interchange, east to Lake Henshaw. Periodically, speakers will provide insight and expertise. More intimate and up-close looks will be had on a second day, with visits to the private archaeological and mineralogical treasures at the Molpa archaeology site and Oceanview mine, respectively.

The second part of this guidebook is a collection of papers on a cross section of subjects. These papers are roughly presented based on the *geologic age* of the subject matter within. Therefore, we begin the volume with Bill Elliott's "sketchy" but comprehensive outline of the rocks in a regional setting, followed closely by the pegmatites representative of the latter stage batholitic emplacement. Quaternary processes, followed by stories from the "anthropocene" close out the volume. SDAG is proud to present the first publication of the landmark *Bison latifrons* fossil discovery. We also include updated, abridged or excerpted versions of previously published works that are relevant and illustrative of the geoscience and natural history of the area. For all the content presented, ranging from watermelon tourmalines, to ancient landslides, to petroglyphs, to rainbow trout, interested readers are encouraged to seek out more information via the references provided in any of the included papers.

The third part consists of a generalized geologic map, oversized to adequately present the disposition of the more significant map units across the 50+ mile long drainage area. The elongate east-to-west orientation of San Luis Rey River and Palomar Mountain lend themselves to such a "landscaped" presentation.

Introduction and Acknowledgments

Welcome to San Diego Association of Geologists' 2013 annual field trip. This year we will explore what I now consider to be the "heart" of northern San Diego County, the San Luis Rey (SLR) River valley. Depending on how you count, our county can be divided into nine or ten drainage basins or watersheds. The 562 square mile San Luis Rey basin is one of the largest and least developed, with over 80% dedicated to agricultural use, parks, and open space. Unlike the Santa Margarita River drainage (which we must share with Orange County and the United States Navy), or the Tijuana Watershed (75% of which lies south of the international border), the San Luis Rey River is entirely "owned and operated" by residents of San Diego County. We can take pride in that. As such, the area is deserving of a volume dedicated to the unique geology and natural history found within its limits.

Southern California geoscientists not living under their respective rocks are well aware of some of the unique character of the SLR. The pegmatite districts of Pala, Rincon, and Mesa Grande have been known since the turn of the prior century, and remain curious "pockets" in the Peninsular Ranges batholith; Palomar Mountain is home to a 6000 foot high state park, but more of a ridge than a mountain, thanks to our tectonic regime and the trend of the Elsinore fault system. And for commuters, Highway 76 has always been a preferable alternative to congested Highway 78, and a convenient cutoff for "desert rats" returning from Anza-Borrego Desert and beyond.

This volume strives to "take the road less traveled" and look at the San Luis Rey River valley from some less common angles. To those that know the geology of the Peninsular Ranges batholith like the back of their sunburnt hands, we offer articles on paleontology, archeology, and history. To San Diego history buffs, we present an opportunity go underground into a pegmatite vein, for a look at where it all began.

It is probably no coincidence that our road log, "Bones, Vugs, and Harmony," takes place between Interstate 15 and Highway 79. To the west, the Tertiary rocks underlying the populated Oceanside to Bonsall sprawl, have been well studied and presented, even as recently as last year, with SDAG's *Waiting for Tsunami: Coastal Hazards of Northern San Diego County, California*. And to the east, the Elsinore fault to Anza-Borrego area has always afforded excellent geologic exposures, as presented in SDAG's 2003 and 2010 volumes, the respective *Geology of the Elsinore Fault Zone*, and *Geology and Lore of the Northern Anza-Borrego Desert Region*. By design, this volume stops short of any papers or presentations of the Elsinore fault, as we will leave some cutting edge tectonic discussion to next year's field trip, tentatively planned to cover southern California's complex transform fault system.

Indeed, some generous and knowledgeable folks have contributed their time and effort to help me bring this publication on the central "heart" of the

San Luis Rey to print. When selecting the study area, I made sure and chose an area of San Diego with a large landslide, so as to not exclude my friend, Mr. Mike Hart, from the list of potential contributors. And I must also thank fellow engineering geologist Mr. Bill Elliott, probably the most prolific contributor in the history of SDAG, for giving me more than I could use, but exactly what I needed, in the form of a geologic sketch and longitudinal profile of the River. A big thank you to Mr. Chuck Houser, for dusting off his master's thesis and sharing in a vision that it, along with a reprint of Jesse Fisher's seminal article, for giving this book a gem-like quality. Chuck deserves all credit for serving as the adit through which we access the Pala pegmatites!

From a paleontological perspective, this volume has a huge (Great Plains Bison-sized) leg to stand on thanks to the article on the Bison latifrons discovery by Tom Deméré and the others at San Diego Natural History Museum. I also very much appreciate Norrie Robbins, Shasta Gaughen, and Paul Davis for taking us into the "anthropocene," presenting Native American concepts from a geological and mineralogical perspective. Who knew the original inhabitants of the valley could tell a gabbro from a tonalite, and dug pegmatite veins like the rest of us?

As an engineering geologist, who makes a living assuring our infrastructure is on a firm foundation, I appreciate that with 80% of the valley being unimproved open space, there are still geotechnical and environmental-geology investigations to present. Greg Farrand got a late start but managed to deliver with an informative paper on slope stability along the length of the Escondido Canal.

Behind the scenes, I would like to thank Dave Bloom, Monte Murbach, Greg Cranham, Jim Ashby, and Lowell Lindsay for donating enough of their time and expertise, essentially "making sure I didn't ---- this up." They, and other committee contributors, deserve more credit for this volume than they would probably be comfortable taking. And there have been others, such as Chuck Mauvais, who along with David Kajtaniak, gave me a map to help us all get around. Thanks go to SDSU professor Gary Girty, who promptly responded to my emails and luckily shares my interest in "saprock." Thanks also to Woody Higdon, who never fails to impress with his aerial imagery.

It almost goes without saying, but I should also point out that the San Diego Association of Geologists Fall Field Trip would not take place without generous contributions from our sponsors. These include those listed on the following page, that have provided charitable financial contributions, but also those that have contributed in other less tangible ways. My employer, Fisher and Strickler Rock Engineering, LLC, deserves thanks for not only making the financial contribution that enabled the color printing found throughout this book, but was also amicable to me spending the otherwise "billable hours" necessary to compile this field trip and road log. And in the same category, my understanding wife Dani and son Tek, who (most of the time anyway), understood when I said I had work to do.

Lastly, I would like to thank the first contributor, author and trout fisherman Bruce Campbell. Way back in March of 2013, I accepted your paper with a vision that an article on native rainbow trout would somehow fit into a geological field trip road log. It turned out to be a perfect homage to whom we dedicate this book, fisherman, outdoorsman, and SDAG founder Charles "Mike" Bemis.

Brian J. Olson CEG
Fisher & Strickler Rock Engineering, LLC.
SDAG 2013 Vice President
Field Trip Chair

Dedication

This Volume is dedicated to the Founder of the San Diego Association of Geologists, Charles "Mike" Bemis, shown here with his bird dog "Rhett." Stonewall Peak and Lake Cuyamaca in the background.

Charles G. "Mike" Bemis (1940–2013) attended San Diego State University (BS Geology, 1963). He worked for 50 years as a geotechnical professional, domestically and overseas, first with California Division of Highways, then in private industry. Overseas tours included Iran, Saudi Arabia, and Micronesia. His last 24 years were spent at Ninyo & Moore, where he was Principal Geologist and Director of Business Development.

Mike was co-founder and first president of the San Diego Association of Geologists. He received the 2013 SDSU Geological Sciences Baylor Brooks Alumni Award, and the American Public Works Association (APWA) 2013 Private Sector Award. He served on many boards of nonprofit organizations, including APWA, East County YMCA, and Lake Cuyamaca Foundation. He mentored many young geologists and geotechnical support staff.

Mike's specific geologic interests and professional expertise focused on soil and slope stability, bridge, quarry and dam siting, coastal processes, Pleistocene geology, and ancient and modern sea level changes. As a geologist and as a sportsman, Mike had an abiding interest in all aspects of natural history, and always had a naturalist's tale to bring home from the field.

As we at the San Diego Association of Geologists celebrate our 42nd year, we hope this volume, which includes an article on native rainbow trout alongside professional papers on pegmatites and engineering geology, serves as a lasting dedication and celebration of the life of our founder and friend Mike.

SDAG 2013 *Field Trip Committee:*
Brian J. Olson, Dave Bloom, Greg Cranham,
Monte & Diane Murbach, Chuck Houser, and Lowell Lindsay.

Sponsors

Diamond Sponsor: $1000

H & P Mobile Geochemistry, Inc.
(800) 834-9888
www.handpmg.com

Hargis + Associates, Inc.
www.hargis.com

Ruby Sponsor: $500

Mission Geoscience, Jim Ashby

Cardno ERI, Steve Zigan

Tetra Tech, Dave Bloom

Fisher & Strickler Rock Engineering, LLC

Chambers Environmental Services, Dan Chambers

Taylor & Francis Group

Emerald Sponsor: $100

Dr. Pat Abbott
Sally & Dennis Avery
Richard W. Berry
Curtis Burdett
Larry Cann
Joe Corones
Greg Cranham
Dr, Neal Driscoll
Earth Consultants International
Eco Rental Solutions
Eggers Environmental, Inc. – Dr. Margaret Egger, CHG.
William J. Elliot, Engineering Geologist
Phil Farquharson, CG-Squared Productions
Greg Farrand
Don Wilson Clark/Katy Freese
Carolyn Glockhoff, Caro-Lion Enterprises
Cari Gomes
Dr. Sarah Gray
Rob Hawk, City of San Diego
Jonas & Associates
Diana & Lowell Lindsay, Sunbelt Publications
Monte Marshall
John Minch
Monte & Diane Murbach, Murbach Geotech
Ninyo & Moore – Jonathan Goodmacher
Brian J. Olson
Pacific Drilling, Co.
Peets Coffee and Tea Store #342, Carlsbad
Les Reed, Geotechnical Exploration Inc. (GEI)
Rigid Lifelines
SEALASKA Environmental Services LLC, Mukesh Mehta, P.G., C.E.G.
Gerald Shiller
Bob Smillie, Terra Costa Consulting Group, Inc.
Scott Snyder
Southwest Geophysics, Inc.
David & Jan Steller
Dr. Anne Sturz
Sue Tanges, Southland Geotechnical Consultants
Terra Costa Consulting Group, Inc.
Mary Walke
www.youtubedownloadersite.com
Carole L. Ziegler

The generosity of these sponsors allows our organization to keep membership dues low, reduce rates for students, produce monthly newsletters, purchase supplies, host the annual field trip, produce our geologic guidebooks, and provide scholarships to local geology students. To contact any of the above sponsors, please visit www.sandiegogeologists.org.

Officers

San Diego Association of Geologists

PRESIDENT: Cari Gomes

VICE PRESIDENT: Brian J. Olson

SECRETARY: Jennifer Morton

TREASURER: Lee Vanderhurst

PUBLICATIONS: Lowell Lindsay

WEBMASTER: Carolyn Glockhoff

San Diego Geological Society, Inc.

PRESIDENT: Dave Bloom

VICE PRESIDENT: Mike Hart

SECRETARY: Diane Murbach

TREASURER: Greg Cranham

DIRECTOR: Lowell Lindsay

DIRECTOR: Monte Murbach

DIRECTOR (EX OFFICIO): Cari Gomes

San Diego Association of Geologists is a program of San Diego Geological Society, Inc., a California 501(c)(3) nonprofit corporation.

San Diego Association of Geologists Publications

1973 *Studies on the Geology and Geologic Hazards of the Greater San Diego Area, California*, Arnold Ross and Robert J. Dowlen, eds.

1974 *Recent Geological and Hydrologic Studies, Eastern San Diego County and Adjacent Areas: a Guidebook*, Michael W. Hart and Robert J. Dowlen, eds.

1975 *Studies on the Geology of Camp Pendleton, and Western San Diego County, California*, Arnold Ross and Robert J. Dowlen, eds.

1977 *Geology of Southwestern San Diego County and Northwestern Baja California*, Gregory T. Farrand, ed.

1978 *Natural History of the Coronado Islands, Baja California, Mexico*, Herman T. Kuper, ed.

1979 *Earthquakes and other Perils, San Diego Region*, Patrick L. Abbott and William J. Elliott, eds. *(GSA field trip guide)*

1981 *Geologic Investigations of the Coastal Plain, San Diego County, California*, Patrick L. Abbott and Shannon O'Dunn, eds.

1982 *Geologic Studies in San Diego*, Patrick L. Abbott, ed.

1985 *On the manner of Deposition of Eocene Strata in Northern San Diego County*, Patrick L. Abbott, ed.

1988 *Landslides in Crystalline Basement Terrain*, James R. Evans, ed.

1989 *The Seismic Risk in the San Diego Region: Special Focus on the Rose Canyon Fault System, Proceedings & SDAG Field Trip*, Glenn Roquemore, Susan Tanges, and Marion Wright, eds.

1990 *Geotechnical Engineering Case Histories in San Diego County*, John H. Hoobs, ed.

1991 *Environmental Perils, San Diego Region,* Patrick L. Abbott and William J. Elliott, eds. *(for a GSA annual meeting)*

1992 *Natural History of the Coronado Islands, Baja California, Mexico* (revisited 1992), Lynn Perry, ed.

1993 *Colorado Desert and Salton Trough Geology*, Joe Corones, ed.

1994 *Geology and Natural History of Camp Pendleton, U.S.M.C. Base, San Diego County, California*, Phillip S. Rosenberg, ed. *(2nd Edition,* 2010*)*.

1995 *Paleontology and Geology of the Western Salton Trough Detachment, Anza-Borrego Desert State Park, California*, Paul Remeika and Anne Sturz, eds.

1996 *Geology and Natural Resources of Coastal San Diego County, California*, Tissa Munasinghe and Phillip Rosenberg, eds.

1997 *Santa Cruz Island: Geology Field Trip Guide*, James R. Boles and Werner Landry, eds.

1998 *Geology and Geothermal Resources of the Imperial and Mexicali Valleys*, Lowell E. Lindsay and William Hample, eds.

1999 *Water for Southern California: Water Resources Development at the Close of the Century*, Gregory T. Cranham, ed.

2000 *Geology and Enology of the Temecula Valley, Riverside County, California*, Barbara B. Birnbaum and Kerry D. Cato, eds.

2001 *Coastal Processes and Engineering Geology of San Diego, California,* Robert C. Stroh, ed.

2003 *Geology of the Elsinore Fault Zone, San Diego Region*, Monte L. Murbach and Michael W. Hart, eds. *(joint publication with SCGS)*

2004 *Mining History and Geology of Joshua Tree National Park*, Margaret R. Eggers, ed.

2005/06 *Geology and History of Southeastern San Diego County, California*, David M. Bloom, Phillip T. Farquharson, and Carole L. Ziegler, eds.

2006 *Mines and Geology of the Randsburg Area, An Historical Gem of the Mojave Desert*, by D.D. Trent.

2010 *Geology and Lore of the Northern Anza-Borrego Desert Region*, Monte L. Murbach and Charles C. Houser, eds. *(joint publication with SCGS).*

2011 *Picacho and the Cargo Muchachos*, Todd Wirths, ed.

2012 *Roadside Geology along Sunrise Highway*, by Michael J. Walawender

2012 *Waiting for Tsunami: Coastal Hazards of Northern San Diego County*, Cari Gomes, ed.

2013 *San Luis Rey on Display, Geoscience in Northern San Diego County*, California, Brian J. Olson, ed.

Geologic Time Scale

EON	ERA	PERIOD (SYSTEM)		EPOCH (SERIES)	Age estimates of boundaries (in mega-annum [Ma] unless otherwise noted)
PHANEROZOIC	Cenozoic	Quaternary (Q)		Holocene	11,700 ±99 yr
				Pleistocene	2.588
		Tertiary (T)	Neogene	Pliocene	5.332 ±0.005
				Miocene	23.03 ±0.05
			Paleogene	Oligocene	33.9 ±0.1
				Eocene	55.8 ±0.2
				Paleocene	65.5 ±0.3
	Mesozoic	Cretaceous (K)		Late	99.6 ±0.9
				Early	145.5 ±4.0
		Jurassic (J)		Late	161.2 ±4.0
				Middle	175.6 ±2.0
				Early	199.6 ±0.6
		Triassic (Ʀ)		Late	228.7 ±2.0
				Middle	245.0 ±1.5
				Early	251.0 ±0.4
	Paleozoic	Permian		Lopingian / Guadalupian / Cisuralian	299.0 ±0.8
		Carboniferous	Pennsylvanian	Late / Middle / Early	318.1 ±1.3
			Mississippian	Late / Middle / Early	359.2 ±2.5
		Devonian		Late / Middle / Early	416.0 ±2.8
		Silurian		Pridoli / Ludlow / Wenlock / Llandovery	443.7 ±1.5
		Ordovician		Late / Middle / Early	488.3 ±1.7
		Cambrian		Late / Middle / Early	542.0 ±1.0
Precambrian					~4600

Note: ka = kilo annum, thousand years ago, Ma = mega annum, million years ago.

Reference: U.S. Geological Survey Geologic Names Committee, 2010, Divisions of geologic time—major chronostratigraphic and geochronologic units: U.S. Geological Survey Fact Sheet 2010–3059, 2 p.

QAP Diagram

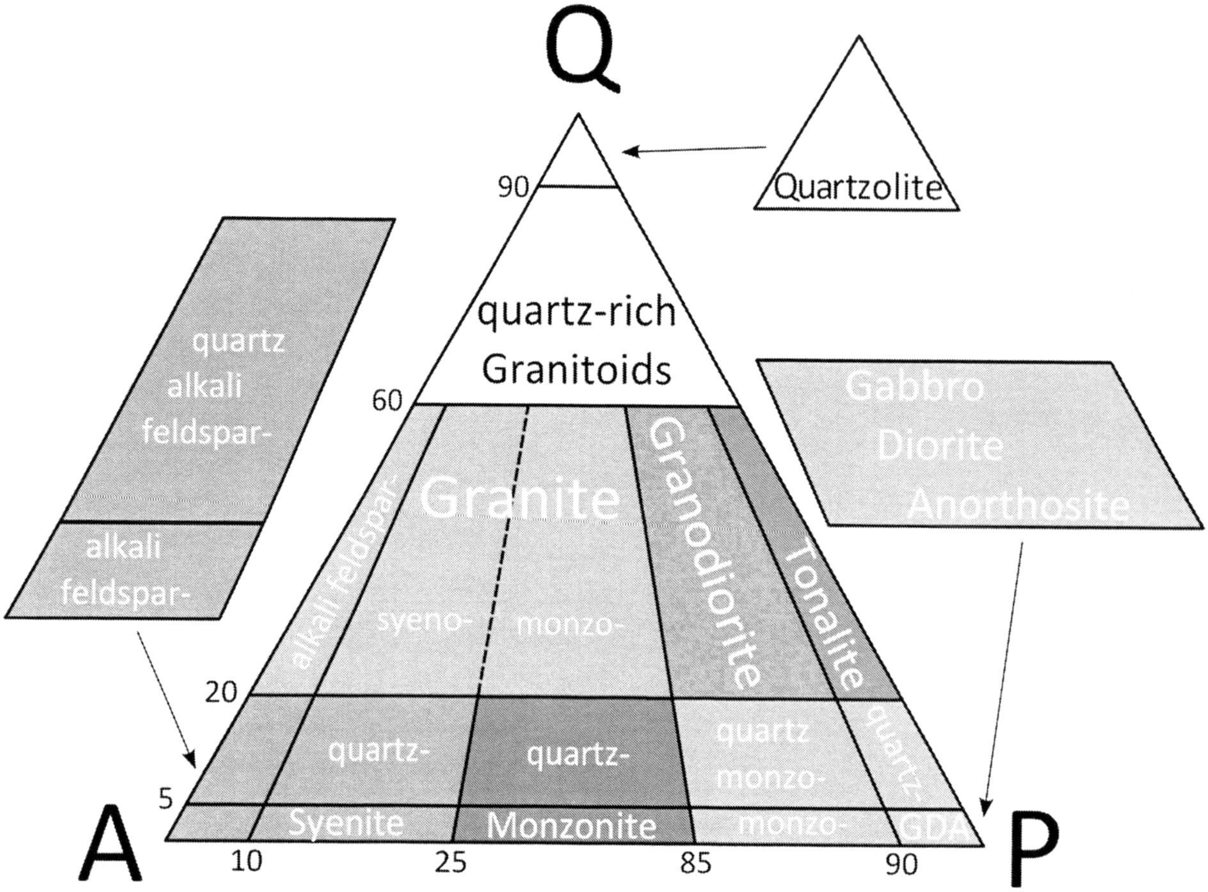

A QAP Ternary diagram is used to classify igneous rocks with visible mineral grains (phaneritic texture). The normalized relative percentages of Quartz (Q), Alkali Feldspar (A) and Plagioclase (P) are used to provide specific name to rocks generally referred to as "granitic" or "granitoid." Based on the QAPF-Diagram from Streckeisen (1974). Only the upper QAP diagram is shown. Modified from a figure published by Wikipedia user Tobias1984 (Wikipedia, 2011). *Permission to use granted by the Creative Commons Attribution-Share Alike 3.0 license. http://creativecommons.org/licenses/by-sa/3.0/deed.en*

Campground and Vicinity Maps

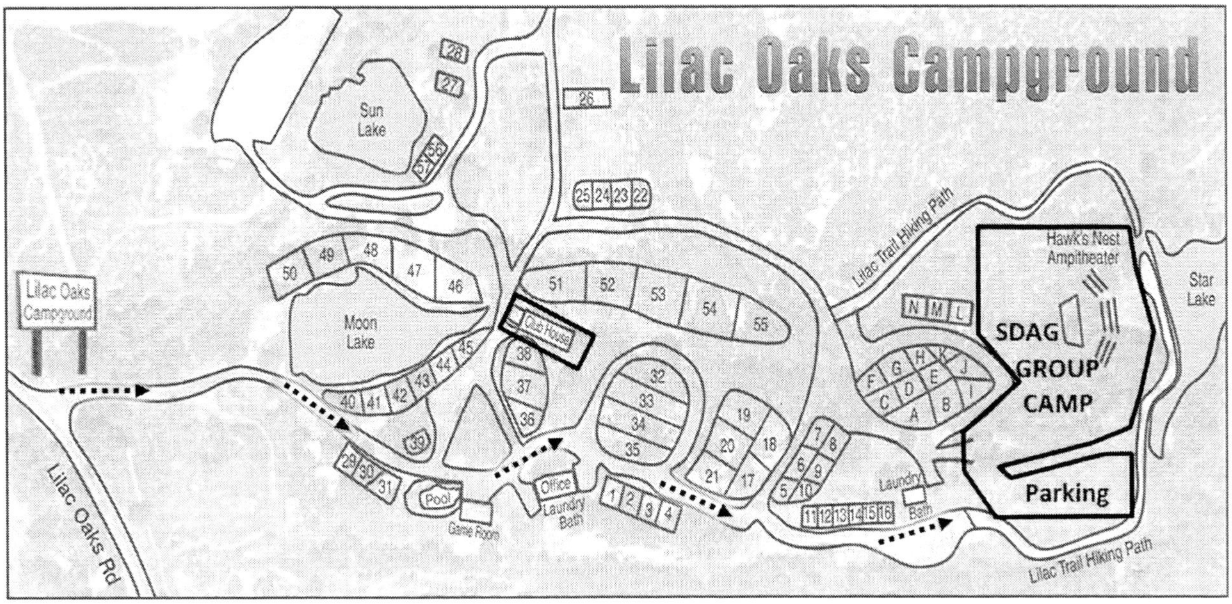

Lilac Oaks Campground
30821 Lilac Road
Valley Center, California 92082

Reference: http://www.lilacoakscampground.com/images/LilacOakswebbrochure.pdf

An some rin up hil and down dale, knapping the chocky stanes to pieces wi' hammers, like sae mony road makers run daft — they say 'tis to see how the warld was made.

—Early description of geologists,
from *St. Ronan's Well*, by Sir Walter Scott, ca. 1820

II. ROAD LOG

PLEASE READ:

This book describes driving and hiking situations and other potentially hazardous activities that are the sole responsibility of the reader. Exploring geological features in mixed terrain from remote wilderness backcountry to relatively urban environments such as the San Luis Rey watershed is not without risk, and all who engage should be aware of their own capabilities, be knowledgeable, prepared, and alert. When exploring such environments, please abide by the laws of the road, respect private property signage, and don't forget about the universally critical use of common sense. The conditions, access and features described in this book can and do change with time.

Previous page: *The first Palomar "Stages" were horseless carriages, capable of ascending Nate Harrison Grade as early as 1904. Nate Harrison on the running board. (Milton Barly Photo, from Brueggeman, 2008.)*

Bones, Vugs and Harmony: *San Luis Rey River Valley from Interstate 15 to Lake Henshaw*

Convene at the Lilac Oaks Campground, 30821 Lilac Road, Valley Center. The campground overflow parking area is located on a dirt soccer field in the southeastern corner of the facility.

Odometer NOTES

0 **Depart the Campground.** Exit from the campground, turning right. Zero the trip odometer as you turn onto Lilac Road.

The east-west drainage through the campground follows roughly the contact between mid-Cretaceous monzogranite of Merriam Mountain to the south and tonalite of Couser Canyon to the north. Most of the campground is located with what's mapped as tonalite of Couser Canyon, a hornblende-biotite tonalite, containing some granodiorite and is characterized by an abundance of pegmatitic dikes. Our campsite, located at the eastern end of the campground appears to be just south of the contact in monzogranite of Merriam Mountain, a medium to coarse grained, massive leucocratic hornblende-biotite monzogranite (Kennedy 2000).

0.3 **Turn Left onto West Lilac Road.** Follow this road due west for nearly two miles. The geologic map of the Pala 7.5-degree quadrangle (Kennedy 2000) shows the road roughly following a contact between Cretaceous granitic units. To the south, is monzogranite of Merriam Mountain. Rocks north of the road are tonalite of Couser Canyon. At about mile 1.2, we cross the north-trending Lilac Tunnel of the San Diego Aqueduct. We cross over this aqueduct again as we head up San Luis Rey River watershed between stops 1 and 2.

2.1 **Turn right to remain on West Lilac Road.** Travel north to northwest. The road winds among some well-kept residential properties, small agricultural enterprises, and shallow tonalite outcrops.

4.5–5.5 As the road generally straightens and assumes a more westward trend, we travel out of the tonalite of Couser Canyon, into massive, medium-grained, leucocratic biotite granodiorite, mapped as granodiorite of Indian Mountain by Kennedy and Tan (2007). To the north, or right hand side, glimpses of corestones and precariously balanced rocks can be seen across the landscape, a weathering habit not unique but seemingly characteristic of the "monzogranites" in this area of San Diego County. The canyon to the right is Keys Canyon, a northwest flowing tributary of San Luis Rey River.

5.9 **Cross the Walter F. Maxwell Memorial Bridge.** Many San Diego County residents are used to traveling under this Lilac Road or "rainbow" bridge, on the way in or out of the county on Interstate 15. The Lilac Road Bridge is a reinforced concrete arch bridge, built in 1978. It spans 695 feet and was designed by Caltrans architect William Wells.

According to Caltrans (2013), Walter F. Maxwell (1909–1980) was a civil engineer who built over 570 bridges and structures in southern California. After World War II, in which Mr. Maxwell attained the rank of Lieutenant Colonel in the Engineering Corps, he formed the W. F. Maxwell Co., General Engineering Contractors, in Los Angeles and later in Fontana and San Diego. He gained national recognition by serving for six years as co-chairman of the American Association of State Highway Officials-AGC Joint Co-operative Committee.

The Bonsall area was originally called Osgood after California Southern Railroad chief engineer Joseph O. Osgood; the renaming was supposedly an unsuccessful attempt to convince Osgood to build the railroad through the San Luis Rey River valley (the rail line eventually followed the Santa Margarita River instead). Following a contest, the town was renamed Bonsall in 1890 after local minister James A. Bonsall (Fetzer, 2005).

6.1 **Turn right onto Old Highway 395.** The bedrock as we approach San Luis Rey River is mapped as metavolcanic and metasedimentary rocks undivided (Cretaceous and Jurassic) - low grade (greenschist facies) rocks that are in part coeval with and in part older than the Cretaceous plutonic rocks they lie in contact with.

8.4 **After the intersection with State Route 76, turn right into the fenced Park and Ride parking lot.** At this location, there is a Mobil gas station. Use it to gas up and purchase last minute provisions at the market. A full tank at this point is recommended.

Reset the odometer to zero.

STOP 1A *BISON LATIFRONS* DISCOVERY SITE

Speaker: Kesler Randall

Pleistocene fluvial and lacustrine deposits are preserved in patches along the San Luis Rey River valley between Interstate 5 and Interstate 15. Over the past 20 years, a series of residential, commercial, and roadway construction projects along the river valley have resulted in the discovery of "Ice Age" vertebrate fossils including teeth, bones, and partial skeletons of fish, amphibians, non-avian reptiles, birds, and mammals. In April 2013, freeway improvements by Caltrans to the Interstate 15 – State Route 76 interchange near Pala Mesa unearthed a partial skeleton of a bison. Recovered fossils include a nearly complete skull and an articulated series of thoracic, lumbar, and sacral vertebrae, as well as both pelvic bones. An isolated left humerus found 60 feet south of the other bones may be part of the same individual. Preliminary analysis of morphometric skull data suggests a taxonomic assignment to the extinct giant bison, *Bison latifrons*. The fossils were found near the base of a 25 foot thick sequence of Quaternary sedimentary rocks in a light yellowish gray, silty, very fine-grained, micaceous sandstone with calcrete nodules. An irregular erosional contact separated the fossiliferous sandstone from an overlying dark yellowish orange to brown, very fine-grained sandstone. The fine-grained Pleistocene fluvial strata were capped by a yellowish brown, coarse-grained, pebbly sandstone of presumably Holocene age. The age of the Pleistocene strata has not yet been determined, but the occurrence of *Bison* indicates correlation with the Rancholabrean North American Land Mammal Age and a maximum age of about 220 ka. This discovery represents the first record of extinct giant bison in San Diego County and serves to expand our understanding of the paleobiogeographic range of this prehistoric mammal. Previous records of bison from San Diego County have been based on fragmentary remains that are either not identifiable to species level or have been assigned to the more diminutive species, *Bison antiquus*.

STOP 1B MOBIL SERVICE STATION GROUNDWATER INVESTIGATION

Speaker: Todd Wirths

Ongoing environmental investigative activities at the Mobil service station have provided a glimpse into the nature and character of the rock and groundwater conditions that underlie the area. As a result of several monitoring wells installed in and around the site to assess the effects of a previous gasoline leak into groundwater, the existence of a complex fracture-fill aquifer system has emerged within the unweathered tonalitic bedrock.

At least two levels of groundwater exist below the site vicinity, with another shallow level possibly as a result of planter irrigation bordering the property. Additionally, a boring for a proposed well located in the Park and Ride lot was drilled to 65 feet deep with no evidence of groundwater.

For three offsite well borings, digital 360° photography was employed to document the nature of the fracture system encountered during drilling of the well boreholes. Images reveal the vivid petrology of the tonalite bedrock, including crystals, mineral flow bands, xenoliths, and quartz veins. The majority of the mineral assemblage consists of quartz, feldspar, and mafic accessory minerals. Fractures appear as hair-line sinusoidal traces in the boring wall image. A few fractures appear to be filled with quartz, and many are parallel to mineral flow bands. Overall, fracture attitudes and clustering tended to be variable, but within the borings photographed, two predominate sets of fractures were approximately identified: one with a mean fracture dip-azimuth ranging from N020 to N044 with a mean fracture dip plunge ranging from 42° to 52°, the other with the mean fracture dip-azimuth ranging from N128 to N144 and the mean fracture dip plunge ranging from 31° to 54°. However, only the boring for well MW10 (in the Park and Ride lot) was identified with fractures having a mean dip oriented to the southwest; similarly, only well MW-10 produced groundwater at the same level with most onsite wells (roughly 20–22 feet below the surface), suggesting the two parameters may be related (Geotracker, 2013).

0 **Zero mileage if you haven't already done so.** Frequent odometer resets at appropriate places make the road log more useful for those that may desire to join mid-trip. **Exit the Park and Ride to the west. Turn left onto the Old Highway 395, and left again onto State Route 76.**

Aerial view of the Rosemary's Mountain Aggregate Quarry, June, 2011. View North. Photo: Woody Higdon, Geo-Tech Imagery International.

1.6 **Rosemary's Mountain Quarry.** We pass Rosemary's Mountain, Palomar Aggregates Quarry, now being operated Granite Construction. The 36 acre property includes most of the east facing slopes of the Rosemary's "Mountain" (it is probably more appropriately a knob or hill). Although the site has only been active for a year or two, site specific environmental geologic investigations were performed in the late 1980s to mid 1990s (DPLU, 1996). The site is largely underlain by granodiorite, and hand samples averaged approximately 50% quartz, 42% plagioclase, 8% alkali feldspar, with mafic minerals (hornblende, biotite) comprising less than 5% of the rock mass (Bing Yen & Associates, 1990 as reported by DPLU, 1996). A more mafic tonalite was mapped on the north side of the property. It was also noted that P-wave velocity estimates were obtained through seismic refraction as well as lab testing of rock core; velocities of 7,000 to 15,000 feet per second were assigned to the rock between depths of 15 to 60 feet, respectively.

Based on County of San Diego DPLU Notice of Determination (NOD), the operation is projected to continue over a period of 20 years, producing an estimated 202 million tons of concrete and asphalt aggregate (DPLU, 2002). This operation replaces many alluvium mining operations once active up and down San Luis Rey River, providing sand and gravel for northern San Diego County's portland cement concrete and asphalt roadways.

2.0 After a right-hand bend, pull over to right at eucalyptus tree-lined road.

STOP 2 MONSERATE NARROWS

Speaker: Brian J. Olson

This is a relatively narrow passage in the San Luis Rey drainage once referred to as Monserate Narrows. In 1937 the State of California Department of Public Works published Division of Water Resources Bulletin No. 48-A (CDPW-DWR, 1936). This report investigated two possible dam sites: this one as well as another further west at Bonsall. Abutments were proposed at this relatively narrow "bottleneck" in the San Luis Rey, between Rosemary's Mountain to the north, or right abutment, and Lancaster Mountain to the south, or left abutment. The report documented an inundation area that extended from here through Pala, nearly 3 miles east. There is plenty of room to turn off to the south side of the road in front of the gate to Jamie's Lane.

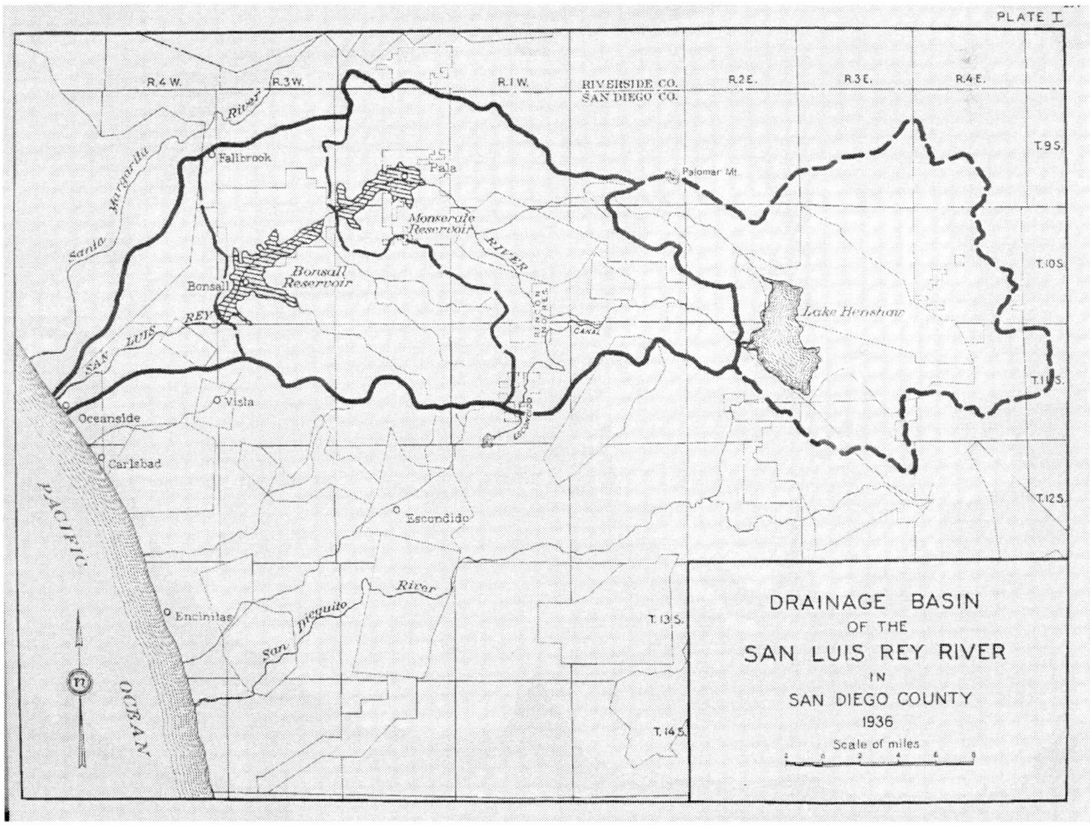

Hypothetical inundation areas are cross-hatched. Source: Division of Water Resources Bulletin No. 48-A.

The largest, highest quality groundwater and surface water flow data measurements also come from this narrows (Robbins, personal communication). Data collection presumably began as early as the mid-1930s as three wells (Numbered 5, 6, and 7) were drilled at the Monserate dam site, presented in the Bulletin No. 48-A. Data collection continues to the present day, presumably relative to the proposed Gregory Canyon landfill we will discuss at the next stop. After this teaser, continue east on Highway 76.

TABLE 11

LOGS OF WELLS DRILLED AT MONSERATE DAM SITE

Depth, in feet	Well No. 5	Depth, in feet	Well No. 6	Depth, in feet	Well No. 7
0	Elev. 283 Sand and silt	0	Elev. 283 Sand and silt	0	Elev. 280 Sand and silt
10	Clay	18	Sand and water gravel	9	River sand & gravel
14	Water gravel	70	Blue clay	45	River gravel
16	Clay with hard streaks growing harder	74	Boulders and clay	60	Black river silt
40	Very hard decomposed granite	78	Seemed to be cemented sand wash gravel and clay.	61	Water gravel
43.5	Hard granite	104	Same with more clay	75	Boulders then into cemented sand, gravel and clay carrying wash gravel, which seemed to be in water bearing strata
45.5	Extra hard granite	106	Hard but still carrying wash gravel which seemed to be in water bearing strata	102	
46.2		122			

Source: Division of Water Resources Bulletin No 48-A.

2.5 **Intersection with Couser Canyon Road.** This road provides a scenic route one may use to return to Lilac Road and Lilac Oaks Campground. It is a narrow road, with twists, turns, and views of bucolic environs north of Valley Center.

4.8 On the right are abandoned dairy farms, evidence of past agricultural use of the valley. The western front of Palomar Mountain is in view on the skyline. The Elsinore fault zone is mapped along the base of the mountain.

5.2 **A grassy lawn opens up to the left.** Look for Pala Rey Youth Camp signage on the right, a weathered white wooden sign. **Turn Right onto the dirt road and travel approximately 0.15 mile to a widened turnaround.** Park. Due south, the topographic wedge carved into the granitic nose is Gregory Canyon, a controversial landfill site.

STOP 3 THE SAN LUIS REY WATERSHED AS NATIVE LAND

Speakers(s): Shasta Gaughen and Norrie Robbins

Here we can discuss the merits of this proposal, effects on the quality of ground water and seepage into the drainage of the San Luis Rey River, and the Pala Tribe's view, past, present, and future. Gregory Mountain, which forms the eastern flank of Gregory Canyon, is home to the spirit Takwish, who balances the forces of life and death in Luiseño cosmology and travels as lightening. At this stop, Gregory Canyon, as well as sacred Medicine Rock, is within clear view to the south.

Return to the highway and continue east on SR 76. To the right, points of interest within this portion of the San Luis Rey are on display. One can still make out the now remnant excavations resulting from the mining of the flood plain deposits. The valley was historically a major source of concrete sand (Weber 1963, p. 228). San Luis Rey River, like most rivers in San Diego County, is dammed upstream, sharply diminishing bed load, so that bed deposits are no longer replenished by natural stream-flow. At three o'clock, toward the southwest, light gray granitic rocks dominate, but darker gray gabbroic rocks can also be seen.

6.3 **Intersection with Pala Mission Road.** Turn left here to divert to historic Pala Mission. Time permitting, it would be a worthy visit for anyone interested in California missions. It includes a bell tower, chapel, and other points of interest.

7.0–7.5 To the north of Pala, the hills are crossed by a network of unimproved roads. These become even more apparent when viewed in aerial or satellite imagery. The roads are largely related to pegmatite gem mining. Very active in the early 1900s, mining of this district continues into the present day. Pegmatite deposits include not only tourmaline, but kunzite spodumene, topaz, morganite beryl, and other semi-precious gems.

8.0 **Note the Magee Road turn veering off to the left.** The road heads northeast and provides access to Pala mining district. We will use this road tomorrow to access Pala Chief mine property, and Oceanview mine, currently open to the public, by appointment only (*www.digforgems.com*). The Gems of Pala store, located just off SR 76 on Magee Road, is open Friday, Saturday, and Sunday, *www.gemsofpala.com*.

Note that the road will begin a climb in elevation, 200 to 300 feet above San Luis Rey River, as we climb the Agua Tibia fan. This is the Pala Conglomerate identified by Ellis and Lee (1919). "Valley fill of a type not common in this area occurs in the valley of the San Luis Rey, in the vicinity of Pala. This material is a conglomeratic mass of boulders and residuum, having a thickness of about 200 feet above and extending to an undetermined depth below the present level of the river. The boulders which make up a large part of this formation are all regular, most of them being prismatic blocks with slightly rounded corners showing that they have been transported only short distances.... This deposit is older than the valley fill which underlies the present valley floors ... but it has not been possible to ascertain definitely the relative ages of these formations."

10.5–10.8 **Agua Tibia and Frey Creeks.** Just past San Diego County Wilderness Gardens Preserve, we cross the Agua Tibia and Frey creeks, tributaries of San Luis Rey River. One can surmise the larger granitic boulders were deposited by debris flows sourced from the north. Noted that as recently as 1987, debris flows have filled local drainages. These drainages have their headwaters near the summit of Palomar Mountain.

We now descend down off the Aqua Tibia alluvial fan. It has been postulated (Jahns, 1954) that the conglomerate comprising the fan could represent a relatively large debris deposit, significant enough to have periodically dammed the San Luis Rey. See article on the longitudinal profile of SLR by W.J. Elliott in this guidebook.

12.5 **After orchards come into view to the northeast, pull over onto road shoulder.** Single file should fit us. Across the road, off in the distance to the east, is a good view of the southernmost slopes of the Palomar Mountains. Just up the road, near call box 76-294, is the Pauma Creek crossing. Note how the drainage appears relatively controlled and tame, as we will be visiting the headwaters up Palomar Mountain a bit later.

STOP 4 TIN CAN FLAT LANDSLIDE OVERVIEW, Exemplary Landsliding in the Peninsular Ranges

Speaker: Mike Hart

Overview of the Tin Can Flat landslide from afar.

Continue east on SR 76.

13.3 (0) **Turn left onto Nate Harrison Grade. Zero odometer here.** Note Nate Harrison Grade route (particularly the upper portion) is most easily accomplished in high clearance vehicles most of the year. Four-wheel drive may be prudent in the event roads are wet. If there is snow, all bets are off.

1.6 **Turn right into the dirt turnout.** Parking is limited. Circle in as far as possible to conserve space. Prepare for a ½ mile hike. Pala Tribal Council, the current landowner, granted permission for our visit. Hiking here without permission is trespassing.

STOP 5 THE TIN CAN FLAT LANDSLIDE EXPOSED

Speaker: Mike Hart

Hike approximately ½ mile east along the dirt roads up Harrison Canyon, to the small reservoir. Note that poison oak, plainly visible during our spring reconnaissance, may be more of a challenge to identify at other times of year. A rattlesnake was also living in the rock riprap adjacent to the reservoir this past spring, possibly there to alert the Pala elders to trespassers. Again, we have permission in writing.

Approximately 1.5 foot high flame structure emanating from the shear zone of the Tin Can Flat landslide, indicative of dewatering and depressurization of the basal rupture zone as the landslide came to rest. Exposure is east of the small reservoir in Harrison Canyon. Photo: Mike Hart

2.5–2.6 **Continue up Nate Harrison Grade.** After a tight left hand switchback, note the road cuts on the east or right-hand side. One must look very closely at these outcrops to see that they are in fact landslide debris, as opposed to rock weathered in situ. At this location, we are less than 1000 feet due northwest of the irrigation pond landmark observed during Stop 5. We are also about 300 feet higher in elevation, so clearly up within the landslide debris we observed at that stop.

3.8–4.2 **After a sharp left turn, drive over the headward landslide contact** (Hart, this guidebook) and not so coincidentally, a mapped strand of the Elsinore fault (Weber, 1963). The road begins an uphill traverse to the northwest. A gneissic or schistose fabric to the rock is notable. The geologic map unit in this area is generally older (undivided) metasedimentary rocks, and until you read otherwise, we will be traversing through rocks entrained within the Elsinore fault zone. More on that later.

5.3 **Vehicles may park along southward spur, or remain on Nate Harrison Grade.**

STOP 6 **BILLY GOAT POINT**

No Speaker – but good historical anecdote on Nate Harrison. In this guidebook, refer to "The First White Man on the Mountain," a biographical sketch on Mr. Nate Harrison, excerpted from *Palomar: From Tepee to Telescope.* This is an informative booklet written by Catherine Wood to "preserve certain facts, stories, and pictures of the days when Palomar was different." Possible lunch spot and real good view of the valley.

Continue up Nate Harrison Grade. It can be noted that along this route, at an undisclosed location, is a side road. This is the entrance to the cabin once owned by Nate Harrison. It is private property, and a protected archeological site. Those interested in history can learn more through the *Nate Harrison Historical Archaeology Project.* San Diego State University Department of Anthropology presents photographs, technical reports and other information at www-rohan.sdsu.edu/~histarch.

Continue winding up Nate Harrison Grade. This nine mile road was pioneered in 1900, up this (relatively) treeless shoulder of the Palomar Mountain. It was known in automobile circles as a good test of speed and endurance for the earliest makes of automobiles. Refer to "Mountain Roads," another *Palomar: From Tepee to Telescope* excerpt (also included in this guidebook).

7.5–8.0 Crest the ridgeline. Those in the driver's side of the vehicles may notice a precipitous topographic descent to the north. These are the aforementioned headwaters of Pauma Creek.

STOP 7 **SOUTHERN CALIFORNIA'S NATIVE RAINBOW TROUT**

Speaker(s): Bruce Campbell and Chuck Houser — not that fishermen give up their secret spots that easily.

9.5 Intersection with Doane Valley Road, providing entrance to Palomar State Park. This road is named after the Doane family, who once owned much of what is now Palomar State Park. Nancy Doane evidently had a mining claim in the area. Entertaining information on Nancy's son George: "Aside from his romantic attitudes, his chief distinction was a flowing beard, which covered most of his face and reached to his waist. He was a well-educated man from the San Francisco Bay region and was a local hotel clerk in San Diego in the early days before he went to Palomar. He was once voted the handsomest man in San Diego, so it is said, in some sort of contest, and his flowing moustaches were a point in his favor" (Brueggeman, 2008).

At this junction is a mapped contact between metasedimentary rocks of Jurassic and Triassic age to the south, and Cretaceous tonalite (undifferentiated) to the north. The metasedimentary rock is described by Kennedy (2006) as mostly quartzofeldspathic schist, pelitic schist, quartzite, and metabreccia; the tonalite is described as mostly massive, medium- to coarse-grained, light-gray, hornblende-biotite tonalite with lesser amounts of granodiorite and monzogranite. For the next roughly six miles the road cuts granitic rock.

From Tepee to Telescope: The lookout station for the Forest Service is situated on Boucher Point, which lies above Nate Harrison Grade. The old Boucher house and orchard lie below and to the west of the lookout. The house is unique in that part of it consists of hand-hewn squared logs, covered inside and out with sawed lumber. There is no spring at the house. Water had to be carried from some distance, but a large cement cistern stored rainwater from the roof. It is said that one spring when the water was low, the cistern was packed with snow to insure a sufficient supply for the summer, and the family had ice water throughout the hot season.

Continue east-southeast on County Road S7.

12.7 (0) South Grade Road (S6) intersection, zero odometer. Here State Park Road becomes East Grade Road. Note the South Grade Road (S6) heads southeast, and is a winding descent back down to Highway 76. Remain on East Grade Road (S7), heading east-southeast toward Lake Henshaw.

1.1 Turn right down driveway to Kica Mik Overlook. Park. Use restroom if necessary.

STOP 8 KICA MIK OVERLOOK

Evidently in early 2010, the Cleveland National Forest Service installed this overlook and interpretive display (elevation 5260 ft). Kica Mik means "View to the South." Provided the weather is clear, this overlook offers a majestic vista of the terrain south of Palomar Mountain. With the displayed photograph of the horizon panorama as a guide, one should be able to make out many of San Diego County's prominent peaks on clearer days.

Roadcut of Saprock weathered from granitic basement at Kika Mik Overlook. Photo: Brian J. Olson, July 2013.

Although views are nice, geologists may be drawn to the road cut north across the highway. This cut is similar to and representative of many of the road cuts along S7 as it winds across the top of the ridgeline, in gradual descent toward Lake Henshaw. The white pegmatite veins are clearly visible. Less distinct are relict core stones, or faint boulder-sized knobs of the rock mass characteristic of weathering. This particular road cut exposes tens of meters thick of saprock, in situ weathered rock in which less than 20% of the weatherable minerals have disintegrated. It is notable that the material falls apart relatively easily with hand pressure, but is not clayey or soil-like. Such a depth of weathering likely did not entirely occur in the Holocene, and there is a growing body of evidence that it points to Pleistocene (>10,000 year old) exhumation and exposure of much of the area between the Elsinore and San Jacinto faults (Girty, personal communication).

The depth of weathering and high relief provide an indication of a relatively young age of uplift and formation of the Palomar Mountain ridge. Deep weathering in San Diego County during latest Cretaceous and Paleogene reduced the area to low relief. The young age of the mountain is inferred because the deeply weathered material would have eroded if the uplift had occurred in earlier times.

Continue east-southeast along East Grade Road. The road roughly parallels the trend of the Elsinore fault zone. It is very much a "zone" in this part of California, generally bound by the Highway 76 in the valley below us, and Warner Springs approximately 8 miles east. Watch for bands of (fault bound?) older metasedimentary and gneissic rocks that are mapped generally underlying the valley bottoms to the left or north. Drivers please keep eyes on the asphalt.

Gregory Pacheco Memorial, at the apex of a turn off from the East Grade Road. Looking closely, decorations include everything from "hot wheels" toys, to a Rocky cassette tape and even an unused ticket for an "An Evening with Yes" at a local amphitheater. Photo: Brian J. Olson (July 2013).

1.9 A small green rectangular road sign indicates the Gregory Pacheco Memorial. To the right, is a loop and scenic overlook.

STOP 9 PACHECO MEMORIAL LOOP

A memorial for 20 year-old firefighter Gregory Pacheco is a reminder of the wildfire dangers that the area residents have dealt with. The La Jolla Fire began late in the evening, September 30, 1999, believed to have been started near the reservation gas station along Highway 76. *http://www.palomarmountain.com/1999fire*. The State Park Road also defines the northern limit of the Poomacha Fire in 2007. To the south, evidence in the form of burned trees can still be seen. This fire was paired with the "Witch Fire" to the south, part of southern California's historic "firestorm" in October and November of 2007.

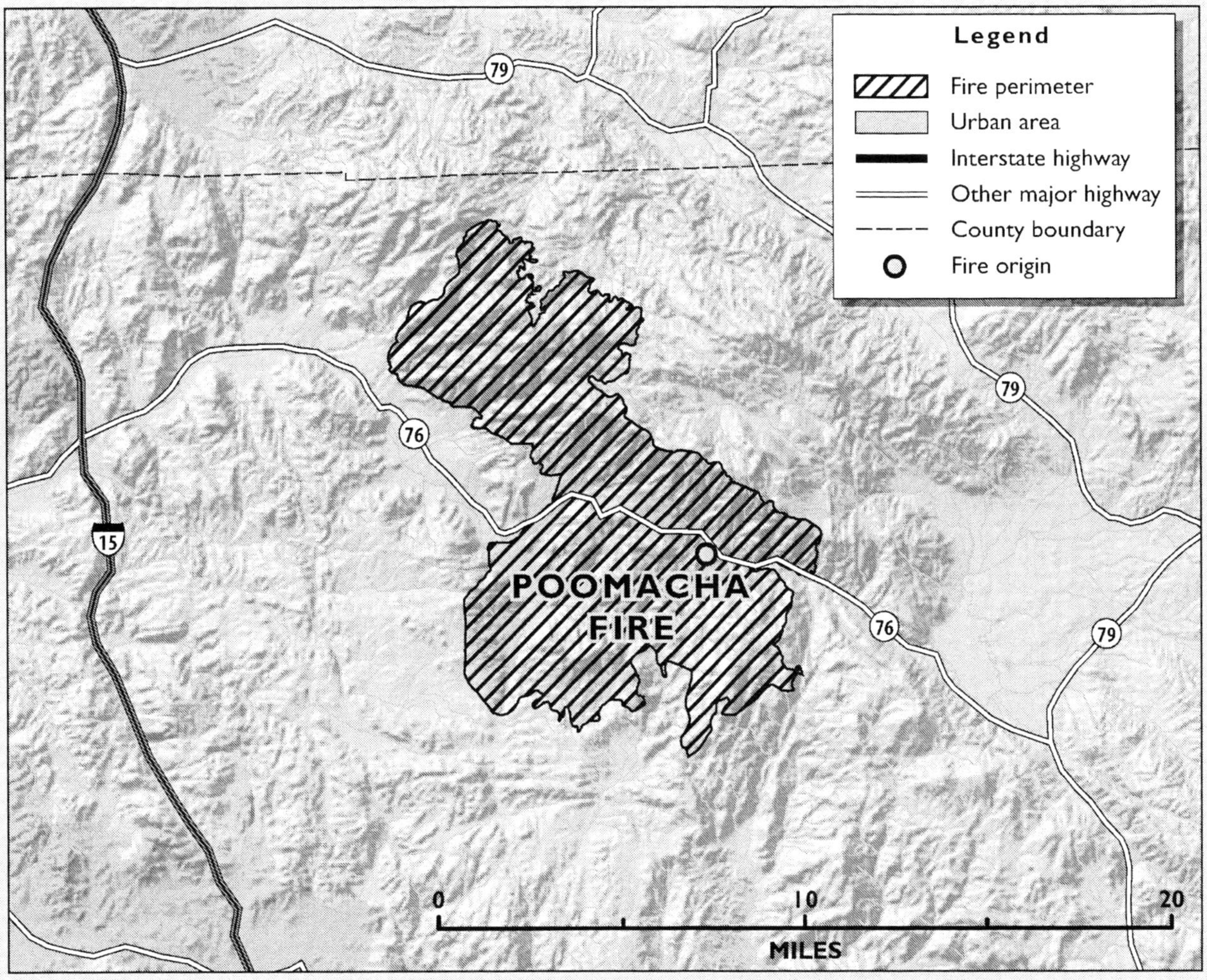

POOMACHA FIRE
Incident#: MVU-010643
County: San Diego
Agency in Command: CAL FIRE, USFS
Start Report Date: October 23, 2007
Containment Date: November 13, 2007
Total Acres: 49,410
Direct Fire Suppression Cost:*
US Forest Service $6,842,500
CAL FIRE $19,368,014
Firefighters Assigned at Peak: 2,793
Structures Destroyed: 217
Structures Damaged: 12
Firefighters Injured: 13
Fatalities: 0
Fuels: Chaparral
Cause: Structure fire

The Poomacha Fire was reported on October 23, at 3:13 a.m., as a structure fire on the La Jolla Indian Reservation, and rapidly spread to the adjacent vegetation.

Reverse 911 was used to contact residents in an effort to evacuate the entire Highway 76 corridor on the first day of the fire. The combined lack of resources and extreme fire weather were so serious on the first day that firefighters could not take action on the fire. It grew from 3,000 acres to 23,000 acres in an hour-and-a-half on day one. Eight injuries were reported.

By the morning of October 24 the area burned had grown to 25,000 acres, and 50 structures had been destroyed. Structure protection remained a priority as several communities, the Palomar Observatory, and a communication tower were threatened.

On October 25, the fire merged with the south side of the Witch Fire.

On October 26, the perimeter control lines of the Poomacha Fire were tied to those for the Witch

Fire. The burn area burned reached 42,000 acres, and containment was at 35%. Extensive damage assessment confirmed 78 structures destroyed. Fire has entered the Aqua Tibia Wilderness and strategies were implemented to reduce impacts. Work in the wilderness is slow due to steep terrain. Residents are allowed to re-enter some areas.

On October 27 the fire progressed to 45,000 acres and was 45% contained. The number of structures destroyed rose to 136, and the number of structures threatened declined from 2,000 to 500. On October 28, the fire perimeter encompasses 49,140 acres. Efforts for the rest of the incident shift toward securing the perimeter, mopping up, and addressing hot spots within the perimeter. Fourteen planes flew the fire on October 28, to contain the spread. It was the first day planes were able to fly in this area. Complete re-entry of residents is expected by evening. By November 1, infrared flights show very little heat left inside the fire perimeter.

The Poomacha Fire was 100% contained on November 13, 2007.

* Figures shown reflect direct suppression cost estimates by stated agency. Other federal, state and local agency costs are not included. Additional expenses not related to direct suppression (rehabilitation, agency overhead, etc.) are not included.

7.8 Pass Mile Marker 3.6, and look for turnout loop to the right.

STOP 10 LOVE VALLEY OVERLOOK

This is an unmarked dirt loop, approximately 200 feet long. Stand on the rock outcrop at the north end of this loop for a good view of Love Valley, a popular hiking destination accessed from the highway farther west. From this vantage, the "perched" nature of the valley is evident. Drainage of the valley has clearly been influenced by the underlying geologic structure, as the area includes two poorly draining or *vernal* ponds, most evident after a wet winter. Ponder the likelihood that this meadow, along with others we've passed in our descent, are remnants of erosional surfaces dating from before Palomar Mountain was uplifted along the Elsinore fault.

Return to the highway. East Grade Road descends toward Lake Henshaw.

10.8–11.1 Parking is inadequate for many vehicles, but slow to catch sufficient glimpse of Henshaw Dam on the left. Sloughy road cuts to the right show evidence of faulting. Passengers may be able to make out west-northwest trending discontinuities, traces that would project under the road, directly at Henshaw Dam. Refer to Photo 9 of Bill Elliott's "Longitudinal Profile of the San Luis Rey," in this volume.

11.3 Intersection with SR 76. Turn Left.

12.1 Pull over into the broad parking area to the right of the road.

STOP 11 LAKE HENSHAW RESORT

Speakers: William J. Elliott and Jennifer Bauer Morton

A general store, easily accessed, serves area visitors, including campers at Lake Henshaw Resort. Stop and stretch, and enjoy refreshments. Generally speaking, the Elsinore fault is not well mapped between here and Riverside County. It is clear that here, the Elsinore fault zone trends northwest-southeast, as does the highway, the southwest side of the lake, and the flank of Palomar Mountain to the north. This is also the gateway to the Himalaya tourmaline mine, the active portion of the Mesa Grande mining district located in the hills to the southwest. The usually locked access road is located at the back of the resort campground.

Across the street, a short walk east affords a good view of Lake Henshaw. The valley floor lies approximately 2700 to 3200 feet (800–1000 meters) above sea level. The network of generally dissected alluvial fans is similar in depositional setting and lithology to those in the Pauba Valley, southeast of Temecula (Todd and Spear *in* Deen et al, 1997). During travel along the Highway 79, which traverses the valley from north to south, road cuts expose indurated sediments similar to the late Pliocene to Pleistocene Temecula Arkose and the overlying late Pleistocene Pauba Formation (Mann, 1955; Kennedy, 1977).

12.9 (0) Zero mileage at intersection with East Grade Road (S7). Return west on Highway 76. We will not re-trace our path, but instead travel west on Highway 76, which follows San Luis Rey River. We will "take the low road."

Northwestward from the S7 intersection, SR 76 closely follows the San Luis Rey River, as well as what can be considered the southern boundary of the Elsinore fault zone. Vegetation is generally well developed, but north side road cuts provide some indication of our alignment with the fault zone.

0.7 Note the corrugated metal guards on each side of the roadway, painted bright white. Here the highway crosses the San Luis Rey River. This crossing, along with another at Old River Road west of Interstate 15, are the only two places the San Luis Rey river is crossed by SR 76.

1.8 It comes up quickly! Just past call box 76-460, turn left to a small picnic area.

STOP 12 PICNIC AREA

Here is good access to the flowing San Luis Rey River, as well as an *intriguing* road cut on the opposite (north) side of the highway. This is a well shaded area, a nice spot for an afternoon snack and refreshment. The road cut opposite the picnic area entrance has been noted (in older references) to display evidence of faulting. More recently, it appears to have been eroded, overgrown, and become generally obscured.

Return to the highway, and continue west through La Jolla Indian Reservation, where the road becomes more curving, following the circuitous San Luis Rey.

6.0–6.1 The river begins to diverge from the highway, and it flows southwest. The highway crosses Cedar Creek, a major southward flowing tributary, and takes us northwest towards Pala.

Approximately 2.5 miles west to southwest of here, Escondido Canal takes off from its connection with San Luis Rey River. G.T. Farrand and H. Van de Vrugt (this guidebook) provide a paper on this canal. This area also coincides with a significant knickpoint in the profile of the San Luis Rey River. Refer to W.J. Elliott's paper, also included herein, and contemplate bedrock resistance, and the steep elevation drop downstream.

6.8 **Near call box 76-410, on the south (left) side, is a short gravel-paved pullout.**

STOP 13 **JURASSIC(?) GNEISS**

Space for vehicles is limited. However, the road cut gravel is noteworthy. The discussion on the next page is extracted from a 1997 Road Log by a pair of "rock docs," Steven Spear and Victoria Todd.

8.7 **Un-named driveway on left.** Speaker Paul Davis resides on a portion of land known as the Molpa Archaeological Site, and has graciously allowed a visit.

STOP 14 **MOLPA ARCHAEOLOGICAL SITE (WINTER VILLAGE).**

Speaker: Paul Davis

The archaeological record of the Luiseño people has been referred to as the San Luis Rey Complex. In fact, this ancient village known as Molpa defines the Type Site for the complex (True et al., 1974). Like all branches of geoscience, new discoveries and interpretations continue to refine the Luiseño's archaeological story, but the site remains a significant and often referenced landmark. See appendix for selected maps and tables from True et al. (1974).

13.0–14.0 The hills to the east (or back over your shoulder to the left if you are heading west) display the distinct white bands of pegmatite that are somewhat less resistant to weathering and erosion as they are predominantly quartz. This is part of the Rincon pegmatite district.

14.6 **Valley Center Road Intersection.** This road heads south and then west to Valley Center, and provides circuitous but convenient access to our campground at Lilac Oaks. Use it as an alternative to Couser Canyon Road that departs SR 76 much farther west.

22.8 **Magee Road. Turn hard right** (or slightly left if you happen to be travelling east). Wind through the orange groves and up the hill for approximately 2 miles, until you reach the gated entrance to Oceanview mine. A "dig host" meets us, collects our paperwork, and escorts us onto the property.

JURASSIC(?) GNEISS

Near La Jolla Indian Reservation, by callbox 76-410, there is a pullout on the left (south) side of the highway. The roadcut exposes gneissic tonalite and granodiorite of Cuyamaca Reservoir (Jurassic?). Lithologically similar plutons in southern San Diego County yielded Late Jurassic U-Pb zircon ages (Miller and others, 1993). This roadcut exposes the westernmost Jurassic pluton in this part of the Peninsular Ranges batholith. To the west, Early Cretaceous western-zone plutons ranging from monzogranite to gabbro were emplaced into Cretaceous and Late Jurassic volcanic-arc rocks (oceanic crust). To the east, Jurassic granitic rocks were surrounded and engulfed by large, mid-to-Late Cretaceous eastern-zone plutons. The western boundary of the belt of Jurassic plutons (I-S line on figure), coincides with prominent gravity and aeromagnetic gradients in the batholith. This boundary, which here lies less than 2 miles west of the highway, apparently marks the fault-controlled edge of the Late Jurassic continent.

The second major Jurassic unit in San Diego County is the Harper Creek Gneiss (HCG), a classic S-type granite that is present in screens on Palomar Mountain to the east. The HCG is peraluminous and carries accessory sillimanite, garnet, and rare cordierite and andalusite. The Cuyamaca Reservoir unit is metaluminous to peraluminous, consisting of two end-members: hypersthene-biotite granodiorite. tonalite ("reduced" series) and hornblende-biotite tonalite. granodiorite ("oxidized" series). These compositional variations reflect the unit's probable origin by mixing of partial melts of metasedimentary rocks (Harper-Creek-type magma) with basaltic magma. Both units contain "foxy" red-brown, reduced biotite, abundant quartz, and ilmenite as the chief opaque oxide reflecting derivation from metasedimentary source rocks rich in organic sulfur and carbon (reduced source).

The light-colored peaks visible to the west are underlain by leucocratic monzogranite and granodiorite called by early workers "Woodson Mountain Granodiorite" after Larsen (1948). In southern San Diego County, such rocks have been mapped as three pluton-forming units, or distinctive magma types, that occur only in the western zone of the batholith; a fourth leucogranite unit occurs only in the central-eastern part. The remarkable lithologic zonation of the PRB reflects systematic changes in source rocks across the batholith resulting from pre-Cretaceous tectonics.

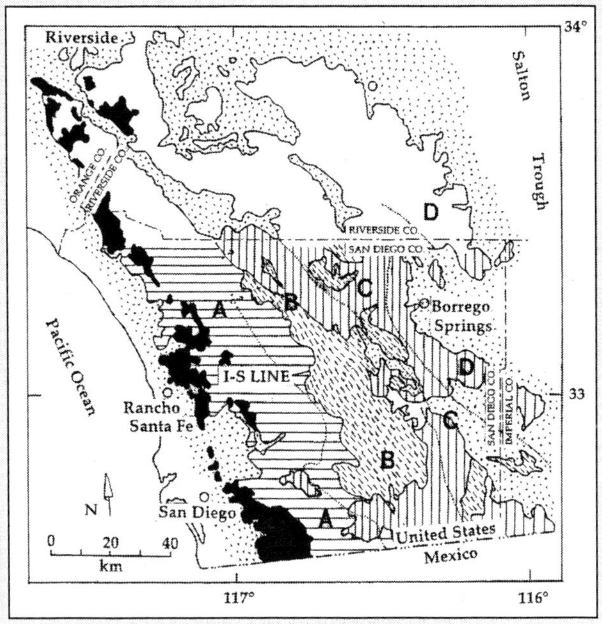

Schematic map of the northern Peninsular Ranges batholith showing volcanic, plutonic, and prebatholithic zones in San Diego County (after Todd and others, 1988). Middle to Late Jurassic Plutons shown by dashed lines; western zone, Early Cretaceous (~125–103 Ma) plutons by horizontal lines (Early Cretaceous plutons that intruded Jurassic belt not shown); eastern zone, mid- to Late Cretaceous (~104-89 Ma) plutons by vertical lines (Jurassic enclaves in eastern-zone plutons shown approximately).

I-S line, western limit of Jurassic granitoids (Todd and Shaw, 1985), coincides with pronounced magnetic and gravity gradients separating western region of residual magnetic and gravity highs and eastern region of sparse magnetic anomalies and gravity lows Jachens and others, 1991).

Dotted lines indicate boundaries between prebatholithic units that occur mainly as metamorphosed screens within batholith; Santiago Peak Volcanics (~130–116 Ma) shown in black; A = Early Cretaceous and Late Jurassic(?) metavolcanic rocks, B = Jurassic-Triassic flysch, C = Triassic(?) to mid-Paleozoic(?) slope and basinal rocks; D = Paleozoic miogeoclinal rocks.

From Todd, V.R. and Spear, S.G., Road Log – Borrego Springs To Pala, in Deen, P.D., Metzler, C.M, and Trujillo, A.P., eds. (1997)

STOP 15 THE OCEANVIEW (ELIZABETH R) PEGMATITE MINE

Speaker: Chuck Houser

The pegmatites of San Diego County are a late-stage formation in the crystallization of the southern California Peninsular Ranges batholith. Pegmatite refers to a coarse-grained (>1–2 cm) igneous rock of granitic composition, typically intruded as dikes. The pegmatites of San Diego County have been mined since the late 1800s for gem tourmaline, topaz, aquamarine, morganite, kunzite, and garnet, as well as a host of accessory minerals. The economic viability of a given pegmatite dike appears to be dependent on several factors including dike structure and chemistry. The Elizabeth R mine in the Pala District serves as model for these factors.

Economic minerals in pegmatite dikes are typically recovered from crystal-filled open spaces referred to as *pockets*. Pockets appear to form in the coarse core zone of the dike where volatiles dissolved in the intruding melt have exsolved and coalesced to form the open spaces. In many mines, this appears to have occurred in areas where the pegmatite dike attains a greater thickness than normal. Thickening of the pegmatite may be controlled by pre-existing weaknesses, such as fracture sets, in the host rock, or by changes in attitude or flexures in the intruding dike. The Himalaya dike in Mesa Grande appears to be a notable exception to this.

The economics of the pegmatite are also controlled by the dike chemistry. Elements such as lithium, beryllium, manganese, and boron, among others, tend to be concentrated in the core zone and pockets of a pegmatite dike. Elements such as these appear largely responsible for the exotic colorations in various gem crystals found in pegmatite pockets. Many pegmatite mines in San Diego County seem to demonstrate the validity of these theories, however, exceptions are abundant.

ROAD LOG REFERENCES AND SOME FURTHER READING:

Brueggeman, P. 2008. *Palomar – From Tepee to Telescope*. 2 ed. Boca Raton: Premier Publishing. 78 p.

Bing Yen & Associates, 1990, Geotechnical Engineering Report of Palomar Aggregates Quarry Site, BYA Project No. 51-151, Prepared for Palomar Aggregates, Inc., c/o Brian F. Mooney Associates, dated October 23, 1990.

California Department of Forestry and Fire Protection (CAL FIRE), U.S Forest Service – Pacific Southwest Region, and Office of Emergency Services (OES), California Fire Siege 2007: An Overview, undated, http://www.fire.ca.gov/fire_protection/downloads/siege/2007/Overview_CompleteFinal.pdf accessed September 8, 2013, 110 p.

California Department of Public Works – Division of Water Resources (CDPW-DWR), 1936, Bulletin 48-A, San Luis Rey River Investigation, 49 p.

California Department of Public Works – Division of Water Resources (CDPW-DWR), 1971, Water Wells in the San Luis Rey River Valley Area, San Diego County, California: Calif. Dept. Water Resources, Bulletin 91-18.

California Department of Transportation (Caltrans), 2013, 2012 Named Freeways, Highways, Structures and Other Appurtenances in California, Division of Research, Innovation, and System Information (DRISI), Office of Highway System Information and Performance, Highway System Engineering Branch, p. 222, January. http://www.dot.ca.gov/hq/tsip/hseb/products/Named_Freeways_Final.pdf accessed September 8, 2013.

Clifford, Harold J., Steven G. Spear, Diane M. Burns, and Guy Tapper. 1996. *Geology of San Diego County: legacy of the land*. San Diego, CA: Sunbelt Publications, 175 p.

Deen, P.D., Metzler, C.M, and Trujillo, A.P. eds., 1997, *Geology and Paleontology of the Anza-Borrego Region, California*, Field Trip Guidebook of the National Association of Geoscience Teachers, Far Western Section, 1997 Spring Field Conference, unpublished.

Ellis, A. J., and Lee, C. H., 1919, Geology and Ground Waters of the Western Part of San Diego County, California: United States Geological Survey Water-Supply Paper 446, 321p., approximate map scale (Pl. 3) 1" = 4 miles.

Fetzer, Leland. 2005. *San Diego County place names, A to Z.* San Diego: Sunbelt Publications. 170 p.

Girty, G.H., 2013, San Diego State Department of Geological Sciences, personal communication, July 30-31.

GeoTracker, 2013, Electronic Data Submittal for Mobil 18-034 (T0607399243), State water Resources Control Board, http://geotracker.waterboards.ca.gov/profile_report.asp?global_id=T0607399243 accessed August 13.

Jachens, R.C., V.R. Todd, D.M. Morton, and A. Griscom, 1991. Constraints on the structural evolution of the Peninsular Ranges batholith, California, from a new aeromagnetic map: Geological Society of America Abstracts with Programs, v. 23, no. 2, p. 38.

Jahns, R. H., ed., 1954, Geology of Southern California: California Division of Mines and Geology, Bulletin 170.

Jahns, R.H., and Wright, L.A., 1951, Gem and Lithium Bearing Pegmatites of the Pala District, San Diego County, California: California Division of Mines and Geology, Special Report 7A, June.

Kennedy, 1977, Recency and Character of Faulting along the Elsinore Fault Zone in Southern Riverside County, California, California Division of Mines and Geology Special Report 131, 12 p.

Kennedy, M.P., and Tan, S.S., 2007, Geologic map of the Oceanside 30' by 60' quadrangle, California: A digital database, California Geological Survey, Regional Geologic Map No. 2, scale 1:100,000. http://conservation.ca.gov/cgs/rghm/rgm/Pages/preliminary_geologic_maps.aspx accessed September 8, 2013.

Kohler, S.L., and Miller, R.V., 1982, Mineral Land Classification: Aggregate Materials in the Western San Diego County Production-Consumption Region, California, California Division of Mines and Geology, Special Report 153, 28 p.

Mann, J.F., Geology of a Portion of the Elsinore Fault Zone, California, California Division of Mines and Geology, Special Report 43, 22 p.

Miller, J., G.H. Girty, and M.S. Girty. 1993. Harper Creek and Cuyamaca Reservoir gneisses, CLMSZ: Late Jurassic plutons of the Peninsular Ranges batholith, Geological Society of American Abstracts with Programs, v. 25, p. 122.

Miller, R.V. 1996, Update of Mineral Land Classification: Aggregate Materials in the Western San Diego County Production Consumption Region. California Division of Mines and Geology, Open File Report 96-04.

Roffers, P.D., and Bedrossian, T.L., 2010, Compilation of Quaternary Surficial Deposits in Southern California, Onshore Portion of the Oceanside 30'x60' Quadrangle: A Project of the Department of Water Resources by the California Geological Survey, California Geological Survey. Special Report 217 (Revised), Plate 11, map scale 1:100,000, http://conservation.ca.gov/cgs/fwgp/Pages/sr217.aspx accessed September 8, 2013.

San Diego County DPLU, 1996, Reclamation Plan (Compliance with Section 2772 of the Surface Mining and Reclamation Act of 1975) for Palomar Aggregates Quarry, Prepared for Palomar Aggregates, Inc., September.

San Diego County DPLU, 2002, Rosemary's Mountain, Palomar Aggregates Quarry (P87-021, RP87-001, Log No. 87-2-13) Notice of Determination – NOD, Received October 15, 2002.

Streckeisen, A. L., 1974, Classification and Nomenclature of Plutonic Rocks. Recommendations of the IUGS Subcommission on the Systematics of Igneous Rocks. Geologische Rundschau. Internationale Zeitschrift für Geologie. Stuttgart. Vol. 63, p. 773-785.

Todd, V.R., B.G. Erskine, and D.M. Morton, 1988. Metamorphic and tectonic evolution of the northern Peninsular Ranges batholith, southern California, *in* Ernst, W.G., ed., Metamorphism and crustal evolution of the western United States (Rubey Volume VII): Englewood Cliffs, New Jersey, Prentice-Hall. p. 894-937.

Todd, V.R., and S.E. Shaw, 1985. S-type granitoids and an I-S line in the Peninsular Ranges batholith, southern California. Geology, v. 13, p. 231-233.

True, D.L., Meighan, C.W., and Crew, H., 1974, Archaeological Investigations at Molpa, San Diego County, California. University of California Publications in Anthropology II, Berkeley.

Trujillo, A.P. ed., 2008, Geoscience Investigation in Northern San Diego County and Beyond: Student-Directed Explorations, Field Trip Guidebook for the National Association of Geoscience Teachers, Far Western Section, 2008 Spring Field Conference, unpublished.

Weber, F.H., Jr., 1963, Mines and Mineral Resources of San Diego County, California, CDMG County Report 3. 309 p., Pl. 1, map scale 1" = 2 miles.

Wikipedia. 2011. Streckeisen Plutonic Rocks Only Top.png, created by Wikipedia user Tobias1984. http://commons.wikimedia.org/wiki/File:Streckeisen_Plutonic_Rocks_Only_Top.png accessed August 9, 2013.

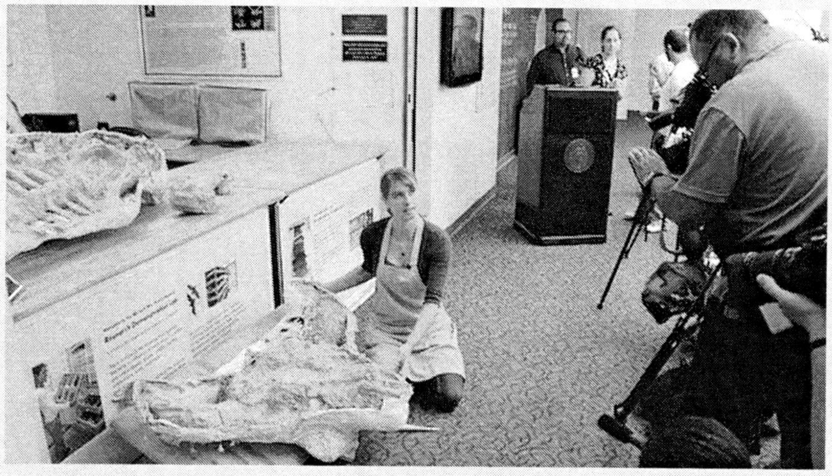

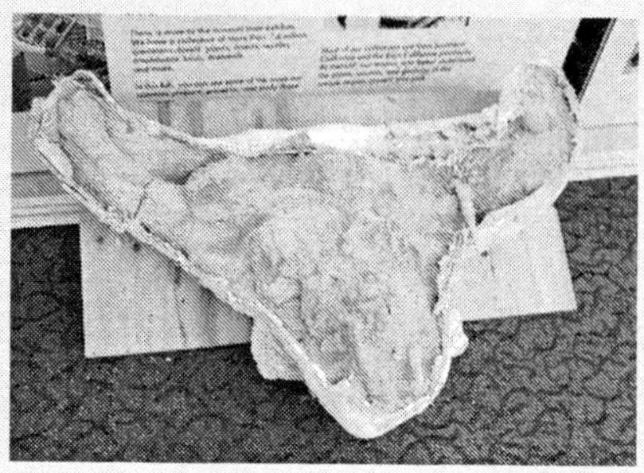

San Diego Natural History Museum shows off a new fossil

Extinct bison bones found during highway construction last month

During construction of a highway ramp in San Diego's North County last month, a paleontologist working for a local museum spotted some ancient bones in the dirt.

Brad Riney, who is described as a "paleontological monitor," found buried treasure for the San Diego Natural History Museum during construction of a new ramp from Highway 76 onto northbound 15, near Bonsall, California.

The alert scientist gathered a crew of five persons to secure the skull and other bones of a huge, prehistoric bison. The museum "tentatively identified" the now-extinct animal as *Bison latifrons*. Officials said the animal roamed the San Luis Rey River Valley between 100,000 and 200,000 years ago, during an Ice Age. The animal once stood eight feet tall at the shoulder and weighed two tons, with the spread of its horns four feet across. The species became extinct about 20,000 years ago, according to museum scientists.

The discovery was made April 10, and the museum crew worked fast to move the fossils from the construction site in just two days. Two bundles of fossils were moved fifty miles south, to the downtown museum on April 12, 2013.

Nikki Anderson was on hand when the museum displayed their new treasure today, May 13, 2013. Anderson said she was one of the workers who carefully removed the bones from the construction site. "Whatever is not fossil, we call *matrix*," the scientist corrected a spectator who first called the environment holding the fossil "dirt."

Nikki Anderson and the other paleontologists first defined a perimeter around the fossils and then soaked strips of burlap in plaster, then layered plaster-strips with wooden splints to create a protected mass that could be safely moved. The bison skull was contained in one block of plaster, and another block was built around the bison pelvis and vertebrae.

Once at the museum, more delicate work is supervised by Dr. Thomas Deméré. The scientists there will use small knives and brushes and handheld "pneumatic scribes" (described as "mini jackhammers") to remove detritus from the bones.

The discovery and preservation efforts did not impact the construction schedule for the highway, according to Laurie Berman, district director for Caltrans, who was at the San Diego Natural History Museum today, May 13, 2013.

Writer Eva Knott for the San Diego READER *website, first published May 13 2013. Photos: Bob Weatherston.*

Photo top right: Nikki Anderson, 30, said the top drawing is the extinct bison, the modern bison is shown below.

Photo bottom right: Fossilized skull of extinct bison still in plaster casing, first displayed today May 13, 2013, at San Diego's Natural History Museum.

III. PAPERS

Previous Page: *San Luis Rey River valley, Monserate Narrows and stream-bottom gravel mining in middle ground. Potential dam site using dark Lancaster Mountain for left abutment and Rosemary Mountain for right abutment. Interstate 15 in distance. Photo: Woodrow L. Higdon, Geo-Tech Imagery International.*

A Brief and Sketchy Geologic History of Northern San Diego County, California

William J. Elliott Consulting Engineering Geologist
P.O. Box 541
Solana Beach, California 92075

Geologic Time

Geology has been described as a six-dimensional science. We are concerned not only with the obvious two dimensional ground surface that we all call home, but we must also strive to understand the third dimension — that which is hidden beneath our feet. Since the earth is literally *older than dirt,* it is important to stretch our minds' boundaries sufficiently to grasp the fourth dimension, the all-important role *geologic time* plays in the assessment of geologic puzzles. To round out the last two dimensions, we must embrace the complex physics and chemistry of a multitude of geologic processes as well as on-going changes that occur with the passage of time. Hence the six dimensions that need to be integrated in one's brain as we ponder and solve scientific problems, as well as engineering issues that affect the health, safety, and lives of citizens who endeavor to make this world a better place in which to work and live.

The concept of geologic time is oftentimes difficult to grasp. The basic idea is that there is lots of it. For example, the current estimate of Earth's age is about 4.6 billion years. Processes that proceed with incredible slowness in terms of human experience can produce profound changes over geologic time. Each rock exposed at Earth's surface is a record of an extraordinarily slow evolutionary process as well as an exceedingly complex history. To unravel the messages contained in each rock seen at the surface of the Earth, it is necessary to consider both how it formed and how it reached the ground surface — two very distinct processes — tasks burdened with interpretation and detail beyond the scope of this essay.

Geologic Time Scale

A copy of the geologic time scale is provided before the road log in this volume. To place a little perspective on the issue of geologic time, Earth's history can be divided into four broad categories.

1. Precambrian (~4600 Ma[1] to ~542 Ma): before life with hard parts that could be easily preserved as fossils.
2. Paleozoic (~542 Ma to ~251 Ma): Age of fishes (and trilobites).
3. Mesozoic (~251 Ma to ~65.5 Ma): Age of reptiles (and dinosaurs).
4. Cenozoic (~65.5 Ma to present): Age of mammals (and humans).

In northwestern San Diego County, and specifically within the San Luis Rey River drainage basin, rocks exposed at the ground surface fall into the last two categories (Mesozoic and Cenozoic) — the last 5½% of Earth's history.

Mesozoic Time

San Diego County has had a relatively short geologic history. The oldest rocks recognized are from the middle of the Mesozoic Era. At that time, from about 170 Ma to 93 Ma, the Farallon plate was sliding steeply and deeply under the North American plate and the San Diego region was quite possibly an island arc, similar to the Japanese islands of today. There was a deep sea trench to the west, and a shallow sea to the east. Effusive volcanoes were depositing great piles of andesitic rock in the

1 Ma is an abbreviation for mega-annum (million years ago).

form of flows and explosive events. This setting is typical of a subduction zone, where one plate of Earth's lithosphere dives beneath another. As it descends into the upper mantle, it partially melts, along with some of the mantle above, producing molten rock that rises up through the crust of the overlying plate to erupt as volcanoes.

Not all of that molten rock, known as magma, reached Earth's surface. In fact, most of it stalled en-route and remained stuck in colder crustal rocks where it cooled and crystallized to form coarsely crystalline granitic rocks that are exposed in San Diego's high back country today. For example, Agua Tibia, Palomar, Vulcan, Rodriguez, and Angel mountains stand as witnesses to these events. While still hot and molten, this melt profoundly altered older rocks into which it intruded.

Several things happened. Older country-rock was partially melted and altered. Some pieces broke off and these xenoliths were incorporated as foreign bodies into the melt. And best of all, chemical alterations occurred in contact zones, sometimes resulting in economic mineralization. In some cases, semi-precious gems were deposited at favorable sites — such as tourmaline and associated gem minerals in the Pala, Rincon, Mesa Grande, and Blue Bell districts within the San Luis Rey River drainage basin.

In other cases, gold was deposited in host schistose country rock (Julian Schist) in the Mesa Grande gold district (Black Eagle and Shenandoah mines) located about 2½ miles east of the Mesa Grande pegmatite district and in the Montezuma gold mining district located about 2½ miles west of Ranchita. Most gold mines in San Diego County were discovered in the 1870s and were mostly played-out or abandoned by early 20th century.

Æons later, tectonic uplift, weathering and erosion during later periods of the Cenozoic Era have exposed these older igneous and metamorphic rocks in eastern San Diego County's backcountry. These batholithic and associated rocks form the backbone of the Peninsular Ranges Provence, which extends from San Jacinto Mountains (south of I-10 at Beaumont and Banning) to well south of the international border in Baja California Norte.

By the end of the Mesozoic Era 65.5 Ma, relative geologic quiet prevailed in San Diego County. Weathering and erosion had worn away earlier mountain ranges, leaving as a legacy, great thicknesses of fossil-bearing sediment in the adjacent Pacific Ocean. A wide variety of invertebrate life of that time is preserved and exposed today in Point Loma's sea cliffs, in sea cliffs along the west side of La Jolla, and at two inland localities northeast of Palomar Airport in eastern Carlsbad and in the foothills of north-central Camp Pendleton. Dinosaur fossils were also recovered from the Carlsbad site.

Early Tertiary

During a long period of early Tertiary weathering and erosion, a thick blanket of decomposed bedrock and surficial pedogenic soils formed at the earth's surface. Most of those soils have been long since removed by erosion. Some notable remnants are described below.

Decomposed Granite. In commercial parlance, "D.G." has been mined from numerous localities in San Diego County over the years. Typically, these strip-mines are sited in deeply weathered, intermediate composition granitic rocks (usually Bonsall or Green Valley tonalite in northwestern San Diego County). Several old quarry sites are located in the San Luis Rey River drainage basin. Most have been abandoned, leveled and redeveloped for commercial uses. Decomposed granite compacts well, is well-drained, and is commonly used for fill, and in soil conditioners.

Core Stones. Prominent rounded boulders, so noticeable in many of San Diego County's foothills and back country are likely also a legacy of that period of early Tertiary weathering and erosion. Chemical weathering breaks down granitic bedrock along three dimensional sets of fractures and joints producing residual core stones surrounded by disintegrated material. Erosion removes these loosened particles, leaving fresh residual rocks as rounded boulders. Commonly, these well-rounded boulders appear to be precariously perched in dramatic fashion on hillsides and mountain tops. This phenomenon, known as spheroidal weathering, can be seen at the proposed Gregory Canyon landfill site, on Palomar Mountain, and other nearby granitic mountains.

Paleosols. The product of intense weathering during ancient periods of warm and humid climate, some of these paleosols can still be seen at several localities in northwestern San Diego County. The closest easily accessible, nearby exposures include: (1) A well-exposed, outcrop at the foot of the westbound on-ramp from Civic Center Drive (old Escondido Avenue) to State Route 78 (Vista, CA), and (2) An almost completely obscured, low cut-slope on the north side of the Sprinter Line railroad tracks, immediately west of Melrose Drive and south of Oceanside Boulevard (south of the ARCO gas station). These varicolored, bleached to warm-colored (yellows, oranges and reds) soils are hard to miss once you've seen them.

Great Rivers. As weathering and erosion continued throughout early Tertiary time, remaining highlands were worn completely away or were buried under some of their own debris. Great rivers with headwaters in Sonora, Mexico, flowed westward during middle Eocene time (46–40 Ma) depositing large deltas of sand and gravel along southern coastal San Diego County. Poway Group Stadium Conglomerate, however, has not been mapped onshore north of San Dieguito River. These distinctive gravels have, however, been found on some of the offshore islands in the Catalina Bight. Coastal lagoons, rich in oysters and fringed with mangroves and nut palms, bordered these deltas.

Islands. Eocene San Diego County was a broad coastal plain with a few scattered hills rising above sea level. Present day coastal foothills such as the San Marcos and Merriam mountains, Morro Hill and Cerro de Calavera likely protruded above sea level as coastal islands in those ancient seas.

Nearshore shallow-marine clastic Eocene sediments (Santiago Formation) were deposited against and immediately westerly from these foothills and nearby islands.

Mid-Tertiary

The Pacific plate, whose eastern edge was following the Farallon plate east across the Pacific Ocean basin, finally touched the North American plate in early Oligocene time, at about 30 Ma. From then on, the Farallon plate continued to subduct at a shallow angle and disappear under the North American plate. Meanwhile, the Pacific plate and the North American plate opted for a different dance and began moving laterally, one past the other, as a right-lateral strike-slip transform fault along the continental margin.

By Miocene time, east-west extension was stretching the western North American continental crust, creating broad arches and swales with accompanying and distinguishing marquis detachment faults in the Basin and Range provinces of eastern California and adjacent western states. Weaknesses thus created in the earth's crust, along with Farallon plate perturbations and probable isostatic adjustments provided the means and conduits for volcanism throughout much of western North America. Two of these intermediate composition volcanic plugs can be visited in northwest San Diego County, one at Morro Hill (Fallbrook), and the other at Cerro de Calavera (Carlsbad).

Coastal Miocene sedimentary strata, a prime example being the San Onofre Breccia exposed in northwestern Oceanside and parts of adjacent Camp Pendleton, include distinctive Catalina schist clasts that could have only come from the west (Catalina and perhaps other islands). These relationships suggest highlands (islands) were present just off shore during this period.

Late Tertiary

At about 6 Ma, the offshore Pacific-North American plate transform fault abandoned its old offshore watery home and jumped east to become the San Andreas fault zone. The result, as we see it today, is the Gulf of California and the Salton Trough (Imperial and Coachella valleys). Right lateral transform movement, at about 2½ inches per year, is moving San Diego County slowly northwestward toward San Francisco County. (At this rate, it will be about 12 Ma before the commute from one county to the other will be short enough to be practical in an all-electric car.) Salton Trough seismicity, geothermal activity (powering *steam* turbine electricity generating plants), and modern Global Positioning System (GPS) measurements provide abundant empirical assurance that the plate boundary is alive, active, and healthy.

With this late Miocene change in tectonic regime, changes also occurred west of the new plate boundary. The old erosion surface was elevated and tilted gently westward at about 1½° (about 2½% grade) creating the Peninsular Ranges as we know them today. High points along the crest include the San Jacinto-Santa Rosa mountains, Palomar, Hot Springs and Vulcan mountains, as well as the Cuyamaca and Laguna mountains. Remnants of the old erosion surface, for example, Warner Basin and Ranchita Valley, are still evident as broad, nearly flat, highland valleys.

A further consequence of moving the Pacific-North American plate boundary inland was a splintering of the continental crust between the old and new plate boundaries. The result is a series of parallel to sub-parallel strike-slip faults between the main San Andreas fault and the western edge of the continental borderland. Most of these splinters are named faults: San Jacinto, Elsinore, Rose Canyon-Newport-Inglewood, Coronado Bank, San Diego Trough, and San Clemente. Interestingly, the amount of lateral plate motion absorbed by each of these faults generally decreases from east to west. In the offshore realm, a series of basins and islands result from oblique movements along these otherwise strike-slip faults. Coronado, San Clemente, Santa Catalina and the Channel islands are present-day members of this horst and graben fault system.

Within the eastern portion of the San Luis Rey River basin, the northwest trending Elsinore fault splinter separates Palomar Mountain and Warner Basin (to the northeast) from Pauma Valley and the Mesa Grande-Rincon gem pegmatite districts (to the southwest). In the early 1920s Henshaw Dam was built directly over the Elsinore fault — at a time when geologic hazards of this nature were either unknown or poorly understood.

Sediments derived from the recently emerged Peninsular Ranges were either washed eastward in the newly formed Salton Trough depression or were carried westward to be deposited in the Pacific Ocean. Some of these sediments are now part of named formations along the coast, for example, the San Mateo Formation (sandstone).

Quaternary

Pleistocene time brought coastal terraces to San Diego County. Between a complex combination of lowered land-levels and raised sea-levels, ocean waves cut broad coastal terraces that today are exposed both north and south of the San Luis Rey River drainage basin.

Along lower elevation, coastal portions of the San Luis Rey River channel, non-marine river terrace deposits can be found where river channels meandered back and forth, cutting new channels and abandoning older channels and associated fluvial deposits in response to changing base levels at its mouth, the Pacific Ocean. Many of these non-marine river terraces can be correlated with nearby coastal marine terraces. Near the intersection of Interstate 15 and State Route 76, *Bison* fossils have been recently found in one such non-marine river terrace. See article in this volume.

The last 2½ Ma of earth history have been dominated by the advance and retreat of alpine and great continental glaciers. Although none ever reached San Diego County, their signature can be seen in geologically young, low-elevation terraces. As ice advanced (taking water from the oceans) and retreated (returning melt water back to the sea), sea levels rose and fell tens to even hundreds of feet. High-stands that lasted long enough, created wave-cut terraces. One of these, created during the ~120 ka[2] Sangamon high-stand (deep-sea oxygen isotope Stage 5e), left sediments (commonly referred to as Bay Point Formation) on a wave-cut platform called Nestor Terrace. Along the northern San Diego County coastline from Del Mar to San Onofre, this thin deposit of silty and sandy debris is relatively easily identified by its drab brown color and intermittent basal gravel. Exposed in upper portions of near-vertical sea cliffs, it unconformably overlies older, lighter colored mid-Tertiary sedimentary strata on both sides of San Luis Rey River.

While sea levels were low during late Pleistocene to early Holocene time, local rivers (for example, the Santa Margarita, San Luis Rey, and Buena Vista) cut deep canyons — only to be back-filled when sea levels rose again. During the past about 6 ka, sea level has been at its present elevation, providing a new temporary base-level for rivers discharging into the Pacific Ocean. Since rivers loose both kinetic and potential energy once they reach sea level, they cannot effectively cut deeper, so aggradation begins. Thus, drowned river mouths are present all along the northern San Diego County coastline. Test borings drilled at and near lagoon mouths typically encounter about 100 feet of Holocene alluvium.

Anthropocene — AKA — *Current Events*

We have all seen signs along major highways identifying prime places for *Wildlife Viewing*. Now here is a new one. *Fossil Seashell Viewing.* From the San Luis Rey River drainage basin, the closest easily viewed sea shells are at Beacon's Surfing Beach, located about 11 miles southerly in the community of Leucadia.

Lagoon restoration is a thinly veiled, fuzzy and misunderstood metonymy for *lagoon remodeling,* or *creation of a manufactured wetland theme park,* as the case may be. A popular pastime in recent years, it really does nothing more than temporarily interrupt Mother Nature's natural and on-going geologic workload to create something more pleasing to those promoting these feel-good efforts. Recent examples of San Diego County lagoon *restoration* projects include San Dieguito and Batiquitos. Yes, the results are beautiful in their own right, but they do not reflect Nature's natural physiographic-hydrologic regime or ecosystem. What do you suppose the odds are that the mouth of San Luis Rey River will fall victim to the drone of earth movers and bulldozers?

What happens next? What of the anthropocene? Only time will tell. To be sure, earthquakes and volcanic activity will continue to occur; prediction, however, will likely remain elusive – at least for the foreseeable future. Mountains will continue to be born, weather and erode, and shed sediment into lakes and seas — very slow processes that are difficult for the average human mind to comprehend. Climate will continue to change, sometimes warmer sometimes cooler, but never exactly the same from day to day or from century to century. The best we can do, as insignificant and fragile human beings, is learn to adjust to whatever Mother Nature puts in our path. Being flexible enough to duck the arrows and roll with the punches, as well as being prepared, will go a long way toward a healthy and prosperous life on this ever changing and dynamic planet. Oh, and by the way, don't leave home without an umbrella and a camera.

No matter what humankind does, Mother Nature will eventually have her way. Everything we do is temporary in the grand scheme of Geologic Time.

Since we all live in the Anthropocene, I encourage you to add your two cents worth to this *current events* section. At the end of this paper there is space for you to jot down your thoughts for future generations to read. If you run out of room, use the margins....

2 ka equals kilo annum (thousand years old)

So, let me leave you with this question. If all humans were suddenly wiped from the face of the earth, how long do you think it would take for every last remnant of human existence (buildings, roads, dams, tunnels and the like) to be eradicated — gone — no trace — done — Earth like it was long before Adam and Eve? There is no preferred answer, but give it some thought, there are some reasonable guestimates. Give up? Try thinking way, way outside the box...

REFERENCES AND SUGGESTED READING

Rather than cluttering this essay with endless citations, it is suggested that the following references be consulted to obtain additional background, detailed descriptions, and expanded insights. For the most part, this list is skewed toward western and northern San Diego County.

Abbott, P. L., 1999, The Rise and Fall of San Diego—150 Million Years of History Recorded in Sedimentary Rocks: Sunbelt Publications, San Diego, California, 231p.

Atwater, T., 1970, Implications of plate Tectonics for the Cenozoic Tectonic Evolution of Western North America: Geological Society of America Bulletin, v. 81, p. 3513–3536.

Ellis, A. J., and Lee, C. H., 1919, Geology and Ground Waters of the Western Part of San Diego County, California: United States Geological Survey Water-Supply Paper 446, 321p., approximate map scale (Pl. 3) 1″ = 4 miles.

Jahns, R. H., ed., 1954, Geology of Southern California: California Division of Mines, Bulletin 170.

Larsen, E. P., Jr., 1948, Batholith and Associated Rocks of Corona, Elsinore, and San Luis Rey Quadrangles, Southern California: Geological Society of America Memoir 29, June 21, 1948, 182 p., map scale 1″=2 miles.

Phillips, R. P., 1983, Geologic History of the San Diego Region: *in* Gomes, C., and Bloom D., eds., Waiting for Tsunami – Coastal Hazards of Northern San Diego County, California: San Diego Association of Geologists, Sunbelt Publications, San Diego, California, 2nd printing of the first edition, p. 5-8.

Rogers, T. H., compiler, 1966, Geologic Map of California, Santa Ana Sheet: California Division of Mines and Geology, Map Sheet, map scale 1:250,000, contour interval 200′.

Thomas, B. E., ed., 1961, Field Trip Guidebook, San Diego County – 1961: Prepared by the Geology Department at San Diego State College (now San Diego State University), March, 1961, for the 57th Annual Meeting, Cordilleran Section, Geological Society of America and Associated Societies, 71p.

United States Geological Survey, 1959-69, Santa Ana, California, 1° x 2° topographic map sheet, N1 11-8, scale 1:250,000, contour interval 200′.

Walawender, M. J., 2000, The Peninsular Ranges — A Geological Guide to San Diego's Back Country: Kendall-Hunt Publishing Company, USA, 114p.

Waring, G. A., and Waring, C. A., 1917, Lavas of Morro Hill and Vicinity, Southern California: The American Journal of Science, Edward S. Dana, ed., Article X, Fourth Series, Volume XLIV, New Haven, Connecticut, p. 98-104.

Weber, H. F. Jr., 1963, Geology and Mineral Resources of San Diego County, California: California Division of Mines and Geology, County Report 3, 309p., Pl. 1, map scale 1″ = 2 miles.

Weber, F. H., Jr., 1982, Recent Slope Failures, Ancient Landslides, and Related geology of the North-Central Coastal Area, San Diego, County, California: California Division of Mines and Geology Open-File Report 82-12-LA, 77p., map scale 1:24:000, contour interval 20′.

Wilson, K. L., 1972, Eocene and Related Geology of a Portion of the San Luis Rey and Encinitas Quadrangles, California [MA Thesis]: University of California, Riverside, Department of Geology, 135 pages, map scale 1:24:000, contour interval 20′.

Wood, S.T., 1974, A Brief History of the Granite Industry in San Diego County: The Journal of San Diego History, San Diego Historical Society Quarterly, Summer 1974, Volume 20, No. 3.

Young, J. M., and Berry, R. W., 1981, Tertiary Lithostratigraphic Variations, Santa Margarita River to Agua Hedionda Lagoon, *in* Abbott, P. L., and O'Dunn, S., eds., Geologic Investigations of the Coastal Plain, San Diego County, California: San Diego Association of Geologists field trip guidebook, April 25, 1981, p. 33-51, geologic map scale 1″ = 0.58 mile.

This essay is inspired by a brief discussion of San Diego County geologic history prepared in 1983 by Dr. Richard Phillips for general public consumption at the San Diego Natural History Museum (Phillips, 1983). Some changes and modifications were made over the years by Dr. Tom Deméré (San Diego Natural History Museum), Dr. Monte Marshall (San Diego State University), Carole Ziegler, William J. Elliott, and Dave Bloom to meet specific audience requirements. It greatly expands the original text and is intended to be consumed by a mature lay audience as well as the professional looking for a quick-and-dirty summary of San Diego County geologic history.

NOTES ON THE ANTHROPOCENE: Your thoughts, ideas, and predictions.

"Daily it is forced home on the mind of the geologist that nothing,
not even the wind that blows,
is so unstable as the level of the crust of this earth."

— *Charles Darwin,* The Voyage of the Beagle

The "Candelabra" Tourmaline. *This specimen was mined by Pala at the Tourmaline Queen mine in 1972. Today it is on public display at the Smithsonian Institution in Washington, DC. Among the many discoveries at the Tourmaline Queen, one stands out — the "Blue Cap Pocket," which was later referred to as the find of the century by Dr. Vincent Manson, then-curator of the American Museum of Natural History.*

The Tourmaline Queen mine located in the Pala district, has always inspired mineral collectors, high graders, and geologists. It was originally claimed by Frank Salmons and Associates in March 1903. Exploratory work yielded some 80 pounds of gem tourmaline. The Queen immediately became the leading producer of tourmaline in the Pala district during the period 1904 through 1914. —Excerpt from "The Queen Reigns Again," by Bill Larson, Lapidary Journal, *1972, www.palagems.com/tourmaline_queen.htm*

Photo: Harold & Erica Van Pelt

Mines and Minerals of the Southern California Pegmatite Province

Jesse Fisher PO Box 77774
San Francisco, California 94107
jef520@comcast.net

Jesse Fisher, who has a master's degree in geology, works in commercial laboratory management. He is also a partner in the UK Mining Ventures specimen recovery project at the Rogerley mine, England. A lifelong mineral collector, he has written numerous articles on the pegmatites of Southern California and the mines and minerals of Northern England. Unless otherwise noted, all photos by Jesse Fisher.

Figure 1. *The Fano mine workings and San Jacinto Mountains, January 1997.*

Figure 2. *Map of the Southern California pegmatite province, including location of major specimen-producing districts. Prepared by William Besse.*

For much of the past century, the Southern California pegmatite province has been the source of some of the world's finest specimens of a variety of gem and rare-element minerals, along with considerable amounts of gem and lapidary material. Although known primarily for producing spectacular specimens of multicolored tourmaline, the region's pegmatites have also been the source of excellent specimens of morganite, aquamarine, topaz, and spessartine, and they were the first recognized occurrence of the lavender-colored gem variety of spodumene now known as kunzite.

Since the beginning of mining in the late nineteenth century, production from the region's mines has been highly cyclic, depending upon the economics of the gem and specimen market. High points of mining activity occurred during the first decade of the twentieth century, and again between the early 1960s to mid-1990s. Although literally hundreds of pegmatites have been discovered and prospected during these periods, only a handful have ever yielded significant amounts

This paper was originally published in a slightly different form in Rocks & Minerals, *v. 86, January/February 2011, p. 14–34.*

of gem-quality minerals. Specimens from the few mines that have, including the Himalaya, Tourmaline Queen, Little Three, Pala Chief, and White Queen, are today the highlight of many collections, both public and private; a few, such as the "Blue Cap" elbaites from the Tourmaline Queen mine, have become iconic.

Geology and Geography

The Southern California pegmatite province is a mountainous, semiarid region located within the Peninsular Ranges of Southern California. Pegmatites in the region are usually clustered into individual districts, which are distributed between the San Jacinto Mountains in central Riverside County southward through San Diego County and into the Baja Peninsula of northwestern Mexico. Much of the region has historically been sparsely populated, though in recent years considerable suburban development has occurred in some areas, particularly near the Pala and Ramona districts.

The majority of the mineralogically complex gem-bearing pegmatites of Southern California have been intruded into rocks of the Cretaceous-age Peninsular Ranges Batholith—formerly referred to as the Southern California Batholith (Larsen 1948). The Peninsular Ranges extend from the area around Mount San Jacinto in central Riverside County to the southern tip of the Baja Peninsula, though the occurrence of complex pegmatites appears limited to the northern section, primarily in Riverside and San Diego counties and the northern part of Baja California (Foord et al. 1991). The granitic rocks that make up the core of the Peninsular Ranges were formed as a result of crustal melting during the subduction of the Farallon plate beneath the North American plate prior to the establishment of the San Andreas fault system, which today forms the boundary between the North American and Pacific plates along much of the western margin of California. The batholith consists of numerous individual plutons, ranging in size from about 10 to 40 kilometers across, and has been divided (Todd and Shaw 1985) into two sections based on chemical, physical, and lithologic differ ences. A younger, more felsic, late/post-tectonic zone, dated to 98–89 Ma, exists to the east; an older, more mafic, syntectonic zone, dated to 135–105 Ma, lies to the west (Kampf, Gochenour, and Clanin 2003). Most of the mineralogically complex pegmatites have been intruded into the San Marcos Gabbro, a mafic unit in the western subsection. Foord (1976) and Snee and Foord (1991) report age dates for the Himalaya pegmatite that place its formation within the time period for the younger felsic zone, and it is likely that the region's pegmatites were created during the emplacement and cooling of these batholiths.

Compositionally, most of the Southern California pegmatites belong to the LCT (lithium and cesium enriched, and with more tantalum than niobium) category, according to the classification proposed by Černý (1991). Many are distinctly enriched in boron, resulting in an abundance of tourmaline in a number of pegmatites throughout the region. Alkali-enriched beryls (morganite and goshenite) are generally more common in districts toward the center of the region, such as Pala, Mesa Grande, and Ramona; aquamarine appears to be more common in pegmatite districts along the northern and southern margins, such as Cahuilla Mountain and Jacumba. Lithium enrichment is commonplace, resulting in the lepidolite present in many pegmatites throughout the region. The occurrence of other Li-bearing minerals such as spodumene is more variable. Spodumene (including kunzite) is abundant in many pegmatites in the Pala district, including the Pala Chief, Vandenberg, Katerina, and Stewart, but is rare or totally absent in pegmatites in the Mesa Grande and Ramona districts. The phosphorous content of pegmatites also appears to be highly variable. Some pegmatites, such as the Stewart in the Pala district and the Pack Rat in Jacumba, have an abundance of both primary and secondary phosphates, whereas others, such as the Himalaya mine in Mesa Grande, have very few. The occurrence of topaz is pretty much restricted to the pegmatites of the Ramona district, Aguanga Mountain, and a few pegmatites in the Chihuahua Valley district.

History

The first reported find of gem tourmaline in California occurred at the Columbia mine on Thomas Mountain in central Riverside County in 1872 (Kunz 1905). It was not until 1898, however, with the discovery of the rich tourmaline deposits at the Himalaya pegmatite in the Mesa Grande district, that any systematic mining for gems began in the region. The Himalaya immediately began producing large quantities of pink and green tourmaline, and when word got out about the discovery, a "gem-mining rush" began, resulting in the discovery of most of the important gem-producing pegmatites in the region over the next few years.

During this early period of activity, much of the region's gem mining was supported by the popularity of pink tourmaline in the Imperial Chinese court of Dowager Empress Tz'u-Hsi. The Chinese demand was principally for carving-grade material rather than true gem-grade tourmaline, making this an important market for material that was not salable as gemstones in the United States. Only pink tourmaline was desirable to the Chinese, and multicolored crystals were usually broken up and the unwanted blue or green portions discarded. Chinese artifacts from this period, such as snuff bottles carved from California tourmaline, can still occasionally be found on the antique market.

The other important gemstones produced by the Southern California mines during this period were kunzite and morganite. Prior to its discovery from a few Pala district pegmatites, the lilac-colored variety of gem spodumene now called kunzite was unknown. Gem spodumene was first discovered by local miner Fred Sickler in 1902, possibly at the Katerina mine. Not knowing what it was, Sickler sent specimens to George Kunz, a gemologist with Tiffany & Co. in New York, who is generally credited with identifying it (though this story is disputed by Sinkankas [1976]). The gemstone—named in Kunz's honor—was quickly popularized by Kunz and Tiffany as a "native" American gemstone. Morganite, previously known as "rose beryl" from a few locations such as Madagascar and Brazil, and referred to as "vorobievite" by Russian mineralogists, had never before been found anywhere in abundance. The name morganite was given to it by Kunz, in 1910, in honor of American banker and mineral collector J. P. Morgan—who also happened to be one of Kunz's best customers.

In 1912 a revolution overthrew the Imperial Chinese aristocracy, putting an end to the Chinese demand for pink tourmaline. Coinciding with that, it appears that the domestic market for California's gemstones had become saturated through Kunz's promotional efforts. These two factors combined to cause a crash in mining activities throughout the region, which only began to revive in the years after World War II.

Most of the material recovered from mining during this early period was used for lapidary purposes; however, a number of superb specimens have survived. Notable among these are the "Steamboat" tourmaline from the Tourmaline King mine, now in the National Museum of Natural History (Smithsonian Institution); the "Postage Stamp" tourmaline from the Tourmaline Queen mine, now in the Houston Museum of Natural Science; and an excellent, if unnamed, cluster of four bicolored tourmalines with quartz and cleavelandite from the Tourmaline Queen mine in the Pala district, which is now in the Harvard University Mineralogical Museum. Sadly, some specimens, originally preserved, have since suffered a less fortunate fate. It is said that prior to the Second World War, the San Diego Natural History Museum had a good collection of minerals from the local pegmatites. When war broke out, the building housing the museum was commandeered for use by the U.S. Navy, and the exhibits were packed away for storage. After the war, the minerals were never found, and it is widely believed that they were dumped at sea with a load of trash!

The renewed prosperity that followed the Great Depression and the war brought with it a rise in interest in lapidary hobbies and mineral collecting and, with these, a renewed interest in local mining. This activity reached a high point between the early 1970s and mid-1990s, centering largely on the most productive mines in the Pala, Mesa Grande, and Ramona districts, along with a few major finds from other, relatively unknown mines in the region. Although more recent mining activities have not reached—and are not ever likely to again reach—the levels of production seen at the beginning of the twentieth century, many spectacular finds of tourmaline, kunzite, morganite, spessartine, and topaz have been made.

This article offers a survey of important finds made between the revival of mining interest in the 1950s and the present. For further information on the geology, mineralogy, and mining history of the region, the reader is referred to the bibliography accompanying this article.

Figure 3 (top left). *Beryl, variety aquamarine, 7.5 cm tall, Fano mine, Cahuilla Mountain, Riverside County. Bryant Harris specimen.*

Figure 4 (top right). *Elbaite with lepidolite and cleavelandite, 11.5 cm tall, Lithia Dike claim, Cahuilla Mountain, Riverside County. Al and Betty Ordway specimen, Jeff Scovil photo.*

Figure 5 (above). *Beryl, variety aquamarine, 6 cm tall, on albite and muscovite, Audrey Lynn claim, Cahuilla Mountain, Riverside County. Recovered by Phil Osborn in 1994. Fisher-Kureczka specimen.*

Cahuilla Mountain District

A number of pegmatites that have produced gem and specimen material are located in the area around Cahuilla Mountain (spelled Coahuila in some older references), in south-central Riverside County. None of these deposits have ever yielded enough material to be commercially viable, but they have been periodically worked by hobbyists and weekend collectors with varying degrees of success. Perhaps the best-known pegmatite mine in this district is the **Fano mine** (also called the Fano-Simmons mine). First worked during the early twentieth-century prospecting rush, Kunz (1905) reports the recovery of kunzite, lepidolite, and colored tourmaline from this deposit. Little evidence of these minerals has turned up in more recent times, and the pegmatite is now known for its somewhat limited production of aquamarine and well-formed crystals of schorl. The best of these were recovered during the 1970s and 1980s by local field collectors, including the late Al Ordway and the Gochenour family.

Another pegmatite of interest in this district is the **Lithia Dike**, located on the southeast flank of the mountain. First discovered by local miner Phil Osborn in the early 1970s, this pegmatite has occasionally yielded some very good pink and green tourmaline. Much of the production from this time was reportedly used for gem and lapidary purposes, but one excellent specimen is an 11-cm-tall pink elbaite crystal on albite and lepidolite in the Ordway collection. The claim was purchased in the early 1990s by collector Paul Geffner, who spent considerable effort trenching the property, only to come away with just a few decent specimens of pink and bicolored tourmaline and a few single morganite crystals (Geffner and Fisher 1997).

The **Audrey Lynn pegmatite**, located northwest of the mountain, has been worked periodically by Osborn since the late 1980s and has produced a limited number of good aquamarine crystals, some large, lustrous schorls, and large smoky quartz crystals. One of the largest cassiterite crystals known from the region, weighing approximately 4 kilograms, was discovered there in 1992.

Chihuahua Valley District

The Chihuahua Valley district is located in a sparsely inhabited area ranging from near Warner Springs in north-central San Diego County northward into southern Riverside County. The best-known mine in this district, and the site of one of the region's more spectacular finds in recent years, is the **Cryo-Genie mine**. The Cryo-Genie, formerly known as the Lost Valley Truck Trail claim, was for many years the site of hobby and weekend digging and would produce the occasional specimen of tourmaline, beryl, or quartz. In the mid-1990s the claim was acquired by Dana Gochenour, and in 2001 he began underground workings on the pegmatite with the assistance of a crew including his brother Ken and local miner Jim Clanin. During the next two years they discovered several major pockets of tourmaline and beryl. The two largest, known as the BAT and Big Monday pockets, yielded numerous large, bright pink tourmaline crystals with a tapering, heavily striated habit. As with most pegmatite finds, many of the larger specimens have been repaired. However, these

Figure 6. *Cahuilla Mountain and the Lithia Dike workings, Riverside County, October 1995.*

Figure 7 (left). *Miner Jim Clanin at the entrance to the Cryo-Genie mine, October 2003.*

Figure 8 (below). *A recently mined tourmaline from the Cryo-Genie mine on display at the Tucson Gem and Mineral Show, February 2002.*

Figure 9 (bottom left). *Elbaite with quartz and cleavelandite, 17 × 14 × 10 cm overall size, Cryo-Genie mine, near Warner Springs, San Diego County. From the Pay Check Pocket, recovered by Dana Gochenour and Jim Clanin in 2003. Fisher-Kureczka specimen.*

Figure 10 (bottom right). *Beryl, variety aquamarine with quartz and cleavelandite, 16 × 12 × 10 cm overall size, Cryo-Genie mine, near Warner Springs, San Diego County. Recovered by Dana Gochenour and Jim Clanin in 2003. Fisher-Kureczka specimen.*

are undoubtedly some of the best specimens to have come from the region in at least the past decade. Perhaps the most spectacular piece from these finds is a large plate of quartz and feldspar with a pink tourmaline approximately 20 cm long projecting from it; the specimen is now in the Los Angeles County Museum of Natural History (LACMNH). A number of additional pockets were found, including the Pay Check Pocket, which produced some very good bicolored pink and green tourmalines, and the Beryl Pocket, which yielded several interesting, tabular aquamarine specimens, many on pegmatite matrix. Unfortunately, the color of the aquamarine crystals proved to be unstable, and some have faded after exposure to sunlight. A detailed account of these finds is given by Kampf, Gochenour, and Clanin (2003).

To the north of the Cryo-Genie, at the eastern end of Chihuahua Valley proper, is the **Blue Lady mine**. The discovery of this pegmatite dates back to the early twentieth century, and although it has never been worked successfully on a commercial basis, it has been the source in recent years of numerous specimens recovered by amateur collectors. The mine is chiefly known for crystals of dark inky-blue tourmaline, often associated with feldspar and quartz, which it has produced in abundance. The mine has occasionally yielded some very good specimens, however. Included among these are a superb 18-cm peach-colored morganite associated with smoky quartz (now in the LACMNH) and what is likely to be one of the best apatite specimens from the region's pegmatites, an 8-cm blue-black tourmaline crystal with a cluster of four tabular, pink apatite crystals perched on one side (now in the collection of deceased local collector Byron Weege).

Figure 11 (top). *Fluorapatite on schorl, 8 cm tall, Blue Lady mine, Chihuahua Valley district, San Diego County. Byron Weege specimen, Jeff Scovil photo.*

Figure 12 (bottom). *Schorl on microcline, 11 × 7 × 6 cm overall size, Blue Lady mine, Chihuahua Valley district, San Diego County. Fisher-Kureczka specimen.*

Figure 13 (above left). *Hydroxylherderite with muscovite and microcline, 7 × 5 × 4 cm overall size, Green Cloud claim, Chihuahua Valley district, San Diego County. Recovered in the late 1990s by Jon Page. Cal Graeber specimen.*

Figure 14 (above right). *Hydroxylherd erite (two generations), 3.5 cm tall, Blue Chihuahua mine, Chihuahua Valley district, Riverside County. Recovered in 1968 by Al Ordway and Josie Scripps. Cal Graeber specimen.*

Also in Chihuahua Valley, the **Green Cloud mine** has produced some of the region's best crystals of hydroxylherderite. Of particular note are a number of large, sharp crystals associated with muscovite and feldspar that were found at the Green Cloud between 1998 and the mid-2000s by local miner/collector Jon Page. The Green Cloud mine was named after the botryoidal, grayish to lime-green mica (assumed to be muscovite) found there. The mine also produced apatite, beryl, smoky quartz, and grayish-green hexagonal tablets of mica. The mine, which is located on private property, has recently been reclaimed to pastureland.

Further to the north, just inside Riverside County, the **Blue Chihuahua (or Blue Dog) mine** was worked during the late 1960s by several local collectors including Al Ordway and Josie Scripps, who recovered some very good hydroxylherderite crystals, as well as some specimens of schorl and topaz. During the past couple of decades, a number of other pegmatites in this area have been prospected by local collectors who have turned up modest amounts of herderite, beryl, and other pegmatite mineral specimens.

Aguanga Mountain District

The Aguanga Mountain district is best known for two mines, the **Ware** and the **Maple Lode**. These mines are situated on opposite sides of a ridge running southwest from Palomar Mountain (site of the famous astronomical observatory) and may, in fact, work different exposures of the same pegmatite.

Figure 15. *The Maple Lode mine, Aguanga Mountain, June 1994.*

The Ware mine (also known as the Mountain Lily, Gem Mine no. 1, and Emeralite no. 2) takes its name from John Wesley Ware who operated the mine and marketed its production, principally blue topaz, through his jewelry store in San Diego in the years prior to the Second World War. More recent attempts during the 1980s to work the mine appear to have been largely unsuccessful, and it is now thought that much of the topaz Ware sold from his shop may have actually been from Brazil. One nice 5-cm single crystal of blue topaz recovered during more recent mining is now in the LACMNH.

Figure 17. *View of Queen Mountain, Pala district, looking from the east, March 2006. The Tourmaline Queen mine is just below the peak at the right; the Stewart mine workings are visible lower on the mountain to the center-left.*

Located not far to the south of the Ware, the Maple Lode mine has been worked with varying degrees of success in recent years by a succession of individuals. Perhaps the most successful were local miners and collectors Roland Reed and Helmut Rohl who recovered a number of specimens of blue topaz, along with some attractive pencil-like crystals of sky-blue tourmaline during the late 1970s and early 1980s. The claim has been recently acquired by local dealer/collectors Cal Graeber and Jeff Kent, who have begun new workings on the property. No major finds have been made as of this writing.

Figure 16. *Topaz, 16 × 12 × 10 cm overall size, Maple Lode mine, Aguanga Mountain, San Diego County. Recovered by Roland Reed ca. 1982. Fisher-Kureczka specimen.*

Pala District

Located in northwestern San Diego County near the village of the same name, the Pala district hosts some of the most famous and productive mines in the region. The district encompasses three adjacent hills, which are host to dozens, if not hundreds, of pegmatites. During the past century, numerous mines have been prospected and developed in the district, but only a handful have actually produced any quantity of high-quality specimens or gem material. A detailed study of the mines and geology of the district can be found in Jahns and Wright (1951).

Queen Mountain, the westernmost of the three hills, hosts several famous mines, including the **Stewart, Tourmaline King,** and **Tourmaline Queen mines**. Both the Tourmaline King and Tourmaline Queen mines were major sources of gem and specimen materials during the early-twentieth-century gem-mining boom in the region. The Stewart mine was active at that time as well but was worked primarily as a source of lithium from large masses of lepidolite; it only came to prominence as a gem-producer during the revival of mining activity in the district in the late 1960s. The reopening of the Queen mine during the early 1970s was famously successful,

Figure 18 (above left). *Elbaite, a doubly terminated crystal 8 cm long, Tourmaline King mine, Pala district, San Diego County. Originally in the Field Museum, Chicago, purchased from George Kunz in 1910. Fisher-Kureczka specimen.*

Figure 19 (above right). *Elbaite with lepidolite, 8 × 5 × 4 cm overall size, Stewart mine, Pala district, San Diego County. Cal Graeber specimen.*

Figure 20 (below). *Entrance to the underground workings at the Stewart mine, June 1997.*

and periodic attempts to work the mine continue. Recent workings at the King mine have been less successful, yielding little in the way of major specimens, and it is thought that early production from the mine came from a single large pocket (Jahns and Wright 1951).

Since its reopening in 1968 by the partnership of Ed Swoboda and Bill Larson, operating as Pala Properties International (PPI), the Stewart mine has been the occasional source of excellent "hot pink" tourmaline, along with lesser amounts of green tourmaline, morganite, goshenite, and kunzite (see Szenics 1970). Because of the broken-up nature of the Stewart pegmatite dike, good-quality specimens, particularly on matrix, have been few, and the majority of the pink tourmaline has been used as lapidary material. The mine was actively worked by PPI from 1968 to 1971 and has since gone through several changes of ownership. It is currently operated by Blue Sheppard, who markets much of what he finds as gemstones and jewelry on a local Southern California cable television shopping channel. Although good specimens from the mine are scarce compared to some other Southern California pegmatites, examples can be seen in both the Smithsonian and the LACMNH as well as in private collections.

Located near the crest

Figure 21 (above). *An exposed tourmaline and morganite-bearing pocket in the Tourmaline Queen mine, ca. 1972. Josie Scripps photo.*

Figure 22 (below). *Beryl, variety morganite, 2.5 cm tall, with lepidolite and quartz, Stewart mine, Pala district, San Diego County. Fisher-Kureczka specimen.*

of Queen Mountain, the Tourmaline Queen mine was reopened in mid-1971 by PPI, who at the end of that year discovered the famous Blue Cap Pocket. A good firsthand description of the find is given by Larson (1972). According to Larson, approximately thirty-three major tourmaline specimens were recovered from this pocket, some of which—such as the "Candelabra," now in the Smithsonian, and the "Rabbit Ears," now in the Houston Natural Science Museum—have become iconic among collectors. Some good morganite specimens have also come from the mine, and a few, such as the PPI Logo specimen, have been combinations with elbaite. The mine continued to be worked by PPI until 1978, during which time several more major tourmaline-bearing pockets were discovered. Since then, several attempts to reopen the mine

Figure 23 (top left). *A watercolor painting of a cluster of elbaite crystals with quartz and cleavelandite, 20 cm across, from the Tourmaline Queen mine, Pala district, San Diego County. The specimen was recovered sometime before 1910 and is now in the Harvard Mineralogical Museum. The painting is thought to have been done by Mrs. W. T. Schaller.*

Figure 24 (top right). *Elbaite with cleavelandite and lepidolite, 10 cm tall, Tourmaline Queen mine, Pala district, San Diego County. Recovered from the Blue Cap Pocket by Pala Properties International in 1972. Fisher-Kureczka specimen.*

Figure 25 (middle left). *Elbaite, 6.5 cm tall, Tourmaline Queen mine, Pala district, San Diego County. Recovered by Pala Properties International ca. 1973. Fisher-Kureczka specimen.*

Figure 26 (middle right). *A couple of miners taking a break outside the Tourmaline Queen mine, ca. 1972. Josie Scripps photo.*

Figure 27 (right). *Beryl, variety morganite with lepidolite, cleavelandite, and elbaite, 7 × 5 × 3 cm overall size, Tourmaline Queen mine, Pala district, San Diego County. Recovered by Pala Properties International ca. 1973. Fisher-Kureczka specimen.*

by PPI in the early 1970s was found by the San Diego Mining Company in 2001, while working the property under agreement with owner Ed Swoboda. For more detailed information on recent work at the Queen mine, see Larson (1972) and Swoboda (2002).

Chief Mountain, in the center of the district, is the site of three productive mines. Most famous is the **Pala Chief mine,** the source of some of the finest gem spodumene (kunzite) crystals found in the entire region. It has also produced some excellent pink and bicolored tourmaline and is one of the few pegmatites in the region that has yielded both. The majority of the finds at the Pala Chief were made in the early part of the twentieth century, and much of the kunzite from this mine was sold to George Kunz and Tiffany & Co. for use in jewelry. A number of important kunzite specimens from this time survive, however, and examples can be seen in both the Houston Museum of Natural Science and the LACMNH. Excellent specimens of tourmaline from this period are in the American Museum of Natural History and the LACMNH.

Attempts to work the mine during the 1970s and 1980s appeared to have had little success. The current owner, Bob Dawson, working the mine on a part-time basis, discovered a pocket in 2001 that yielded several specimens of large

Figure 28 (above left). *Bob Dawson, current owner of the Pala Chief mine, March 2006.*

Figure 29 (above). *Elbaite with quartz, microcline, and cleavelandite, 20.5 cm across, Pala Chief mine, Pala district, San Diego County. Recovered by Bob Dawson in 2001. Bob and Jane Dawson specimen, Mark Mauthner photo.*

Figure 30 (below). *Kunzite crystal, 24 cm tall, Pala Chief mine, Pala district, San Diego County. The specimen was originally owned by George Kunz and then acquired by Martin Ehrmann, who sold it to Harvard Mineralogical Museum in 1933. It was acquired by the Los Angeles County Museum of Natural History in 1987 from Harvard through exchange. Jeff Scovil photo.*

Figure 31 (above). *Former owner Roland Reed inspecting a lepidolite and spodumene-rich section of the Elizabeth R pegmatite, ca. 2002. Brendan Laurs photo.*

Figure 32 (above right). *Beryl, variety morganite, 8.5 × 5.5 × 2 cm overall size, Elizabeth R mine, Pala district, San Diego County. Recovered by Roland Reed in 2005. Fisher-Kureczka specimen.*

Figure 33 (right). *A view of the Elizabeth R (lower right) and Oceanview (upper center) mines, ca. 2002. Brendan Laurs photo.*

Figure 34 (below left). *Beryl, variety morganite and aquamarine, 5 × 4 × 3.5 cm overall size, Oceanview mine, Pala district, San Diego County. Recovered from the 49er Pocket, September 2007. Cal Graeber specimen.*

Figure 35 (below right). *Spodumene, variety kunzite, 4.2 cm across, Oceanview mine, Pala district, San Diego County. Recovered January 2010. Oceanview Mines, LLC specimen, Jeff Scovil photo.*

bicolored tourmaline crystals, along with several kunzite crystals. The best specimen from the pocket consists of a pink tourmaline with a blue termination, approximately 8 cm tall, perched on a matrix of smoky quartz and feldspar. The specimen measures more than 20 cm across and is currently in Dawson's possession.

Nearby the Pala Chief are the **Elizabeth R** and **Oceanview mines.** Little of their early history is recorded, and it is unlikely that they were worked to any great degree during the early period of mining in the region. Both claims were purchased in the early 1970s by local collector/miner Roland Reed, who began development on the Elizabeth R (originally known as the Hazel W) mine. Since that time, Reed has mined the property primarily as a weekend hobby, and he has had several sizeable discoveries of morganite, and more recently kunzite and small amounts of colored tourmaline. Of particular note was a large and impressive specimen of morganite that Reed discovered in 1992. The specimen measured approximately 50 cm across and consisted of several large groups of well-formed morganite crystals to 15 cm on a plate of feldspar matrix. Reed eventually sold the specimen, and it was subsequently broken up into several smaller pieces. The Elizabeth R property was recently purchased by Jeff Swanger, owner of the adjacent Oceanview mine (see below), but it is still being worked by Reed on a part-time basis.

Hiriart Mountain is the easternmost hill in the district and hosts numerous pegmatites and a number of famous mines. Most of the properties on Hiriart Mountain were originally developed by Fred Sickler and were worked primarily for kunzite and morganite. Sickler sold these properties in 1947 to local miner George Ashley, who subsequently sold some while keeping others to work himself. Most important of the Ashley properties were the **Vandenberg** and **Katerina mines.** An often-told story has it that Ashley purchased these mines from Sickler with money made from selling kunzite that he dug from the mines prior to purchasing them. Ashley actively worked both mines during the 1950s and had at least one large find of high-quality kunzite at the Vandenberg in 1954. Both mines were sold by Ashley in the late 1950s, and although little has since been done at the Vandenberg, the Katerina has yielded modest amounts of both kunzite and morganite in more recent years (see Fisher 1999).

Figure 36. *Spodumene, variety kunzite, 5.5 × 3.5 × 1.5 cm overall size, Vandenberg mine, Pala district, San Diego County. Fisher-Kureczka specimen.*

In 2000 Reed sold the Oceanview property to Swanger and Stephen Koonce, who together with a crew including local miner Phil Osborn, began tunneling on the pegmatite. After more than three years of work and 400 feet (121 meters) of tunnel, a major beryl-containing pocket was encountered in September 2007. During the next three months the pocket (named the 49er Pocket) produced numerous specimens of both morganite and aquamarine, many associated with quartz, feldspar, and/or schorl. Although the color of the beryl tends to be pale, the intimate association of morganite and aquamarine, often on the same specimen, is unique for the region. Continued

Figure 37. *Beryl, variety morganite with cleavelandite, 8 × 7 × 3 cm overall size, White Queen mine, Pala district, San Diego County. Recovered in 1990 by Bob Dawson. Joe Ondraka specimen.*

tunneling has recently uncovered a pocket containing some excellent kunzite crystals. A detailed account of the mine and the 49er Pocket is given by Mauthner (2008).

Perhaps the most famous and well-known property on Hiriart Mountain is the **White Queen mine.** The White Queen was among the original Sickler group of claims but saw very little mining during his ownership. After purchasing the claim, Ashley quickly sold it to local mineral dealer Norm Dawson. Dawson did not begin work on the property until 1959 but soon encountered a major pocket of morganite associated with quartz and cleavelandite. Some of the quartz crystals were quite large, and though rather opaque on the outside, when sliced along the *c*-axis and polished, they revealed interesting and delicate inclusions of clay in otherwise transparent quartz.

Several additional morganite-containing pockets were discovered during the 1960s and 1970s. In the late 1980s work at the mine was taken up again by a group including Dawson's sons, Bob and Ken, along with Maury Morgenstein. Beginning in the fall of 1989 a large pocket was excavated, yielding many high-quality morganite specimens. I recall visiting the mine during the summer of 1990 and seeing a shed with many folding tables set up; they were covered with hundreds of morganite specimens, many on cleavelandite matrix. Shortly after that, a dispute over the property with the local Pala Indian tribe resulted in the claim being ruled invalid by the Bureau of Land Management (BLM), and no work has since been done there. The best morganites from the White Queen mine are among the world's finest for the mineral, and perhaps the best known of these is the "Dave Wilber" morganite, now in the Houston Museum of Natural Science.

The Rincon District

The Rincon pegmatite district is located approximately 17 kilometers southeast of the Pala district in the Pauma Valley of north-central San Diego County. The district hosts numerous individual pegmatite dikes, but only a few are known to have produced any amount of gem or specimen materials. The best-known mines in the district are the **Mack, Clark,** and **Victor mines,** all of which were originally prospected during the first decade of the twentieth century. With the exception of some excavations at the Clark mine during World War II for radio-grade quartz, none of these mines has been productive enough to sustain commercial development. Since that time, these three pegmatites have occasionally been worked by amateur collectors with varying degrees of success.

The district is best known for the gemmy, pale blue to bluish-green, prismatic crystals of aquamarine that have occasionally been found at the Mack mine. The Victor mine is best known for producing gem-quality pencil-like crystals of blue to violet-colored tourmaline, but pinks and greens have been found as well, along with a few aquamarine crystals (Sinkankas 1959).

In 1999, local property owner Brian Chuchua, working with Blue Sheppard (Gems of Pala),

Figure 38 (right). *Open-cut workings along the pegmatite at the Himalaya mine, ca. 1905. W. T. Schaller/USGS photo.*

Figure 39 (below). *Main portal at the Himalaya mine, December 1994.*

discovered a helvite-containing pocket on an unnamed pegmatite dike in the vicinity of the Clark mine. Several hundred specimens of well-formed, dark brown, tetrahedral helvite crystals, some more than 2 cm on edge, on a matrix of partially decomposed feldspar and quartz were reportedly recovered from the pocket (B. Sheppard, pers. comm., 1999). Much of the Rincon pegmatite district is on either Indian tribal land or private ranch land, and with the exception of this find, very little mining has occurred in the district for many years.

Mesa Grande District

Although the Mesa Grande district includes a number of mines that have produced gem and specimen materials, it is only the **Himalaya mine** that is of major importance. First discovered in 1898, it was the tourmaline from the Himalaya pegmatite that sparked the gem-mining boom

Figure 40 (top left). *Elbaite, 5.5 cm tall, Himalaya mine, Mesa Grande district, San Diego County. Recovered in the late 1950s by Ralph Potter. Fisher-Kureczka specimen.*

Figure 41 (top center). *Elbaite, 4.5 cm tall, Himalaya mine, Mesa Grande district, San Diego County. Crystal cluster is doubly terminated. Cal Graeber specimen.*

Figure 42 (top right). *Elbaite, 3 cm tall, with quartz and cleavelandite, Himalaya mine, Mesa Grande district, San Diego County. Recovered by Pala International from the Green Pocket in 1993. Fisher-Kureczka specimen.*

Figure 43 (above). *Elbaite with quartz and microcline, 6 × 6 × 5 cm overall size, Himalaya mine, Mesa Grande district, San Diego County. Recovered by Pala International in 1986. Fisher-Kureczka specimen.*

of the following decade. Between its discovery and 1912 the Himalaya Mining Company, headed by J. L. Tannenbaum, produced a considerable amount of tourmaline from the Himalaya mine.

Following the collapse of the local gem market, work at the mine was small scale and sporadic until the property was acquired by Ralph Potter in 1952. During the 1950s Potter succeeded in reopening a number of the old tunnels on the pegmatite and is said to have found several major pockets of tourmaline between 1958 and 1963. In 1963 Potter created the Himalaya Gem Mines Company and sold shares in the mine, retaining partial ownership for himself. Unfortunately, infighting among the shareholders appears to have limited further mining, and in 1968 the main tunnel on the pegmatite collapsed during heavy rains, effectively ending underground work.

In 1977 the mine was leased (and later acquired outright) by PPI. Under the direction of mining manager John McLean, the company initially focused on rehabilitating some of the old workings on the upper portion of the pegmatite. It soon became apparent that lower access was needed to reach untouched ground on the pegmatite, and in 1980 an entirely new adit was driven to intersect the pegmatite low on the east side of the property. Mining was highly successful for the following fifteen years, and numerous tourmaline-bearing

pockets were discovered. Some of the more prolific finds include the President's Day Pocket of 1989, the Green Pocket of 1993, and the Rainbow Pocket of 1996. Details of these finds are available in Fisher, Foord, and Bricker (1998). After two years of driving barren tunnel, the mine was closed and leased to new operators. Shortly afterward, the main tunnel collapsed, limiting further development at the mine.

The Himalaya pegmatite has been very productive for both colored tourmaline and several other pegmatite minerals. Although not known for large matrix specimens such as those found at the mines in Pala, the Himalaya has produced many thousands of tourmaline crystals. Typically, they are doubly terminated and may contain three or more distinct color sections. Of particular note are the largely flawless "gem pencil" crystals, often showing three or more color zones, sometimes found in pockets. While not as common, excellent matrix specimens have been found, typically associated with smoky quartz, lepidolite, microcline, and/or cleavelandite. A good cluster of multicolored tourmaline crystals associated with smoky quartz, dating from the Potter era, is in the LACNHM, and a number of excellent specimens of more recent discovery can be seen in the Houston Museum of Natural Science. Aside from tourmaline, the Himalaya mine has produced excellent specimens of microcline, cleavelandite, lepidolite, smoky quartz, fluorapatite, stibiotantalite, hambergite, and stilbite.

Figure 44 (above left). *Lepidolite with quartz, 9 × 7 × 6 cm overall size, Himalaya mine, Mesa Grande district, San Diego County. Fisher-Kureczka specimen.*

Figure 45 (above right). *Stibiotantalite with elbaite, 3.5 cm tall, Himalaya mine, Mesa Grande district, San Diego County. Recovered by Pala International in 1989. Fisher-Kureczka specimen.*

Ramona District

The Ramona pegmatite district is located in central San Diego County, near the town of the same name. Several pegmatites in the district have historically produced specimens, but work in recent years has focused almost solely on the **Little Three** and the adjacent **Hercules mines**. The mines in Ramona were first worked between 1903 and 1912, and Kunz (1905) reports the production of blue topaz, spessartine, and various colors of beryl. Little mining appears to have happened in the district until the Little Three mine was acquired by Louis Spaulding Sr. in 1951, and mining on the property was resumed a few years later. Upon his death in 1973, mining was taken over by his son, Louis Spaulding Jr., who continued work through the 1990s. The largest find at the Little Three in

Figure 46 (top). *Surface workings on the main dike of the Little Three mine, Ramona district, March 1996.*

Figure 47 (middle). *Elbaite, 5.5 cm tall, with quartz and cleavelandite, Main dike, Little Three mine, Ramona district, San Diego County. Recovered in 1976 by Lou Spaulding Jr. Fisher-Kureczka specimen.*

Figure 48 (bottom). *Topaz, 6 cm tall, Main dike, Little Three mine, Ramona district, San Diego County. Recovered in 1976 by Lou Spaulding Jr. Cal Graeber specimen.*

recent years was the New Spaulding Pocket, encountered in 1976. The pocket yielded numerous specimens of blue topaz, dark green elbaite, and clusters of large lepidolite crystals. Perhaps the most significant specimen from the pocket is a 25-cm plate of cleavelandite and smoky quartz with two large, dark green tourmaline crystals in the center and two blue topaz crystals to one side. This specimen is now in the Carnegie Museum of Natural History. Additional topaz- and tourmaline-bearing pockets were found in 1986 and 1991. A detailed description of the New Spaulding Pocket can be found in Stern et al. (1986).

Figure 49 (left). *Spessartine with cleavelandite and schorl, 6 × 5 × 3 cm overall size, Hercules dike, Little Three mine, Ramona district, San Diego County. Bryant Harris specimen.*

Figure 50 (below). *Open-cut workings on the Hercules pegmatite, Little Three mine, Ramona district, March 1996.*

Just to the northeast of the Little Three is the **Hercules** (also called the **Hercules-Spessartine**) **mine.** Although a separate claim, the Hercules has also been worked in recent years by Spaulding under agreement with the claim owners, Buzz Gray and Bryant Harris. During the 1970s and 1980s, Spaulding alternated work between the two mines and had a number of major finds at the Hercules.

Much of the spessartine that was found consisted of etched or corroded crystal fragments, which were used for gem rough by the claim owners. A number of good spessartine specimens, many associated with albite and schorl, were also recovered during this period, and good examples can be seen in the Smithsonian and LACMNH collections. The best specimens, of which I am aware, however, are in the private collection of Harris.

Beryl, both aquamarine and morganite, has occasionally been found at the two mines, though perhaps the best specimens came from a find on the Hercules dike in 1983. The beryl crystals, both matrix and single crystals, were pink in color (typical of morganite), but they had a prismatic (rather than tabular) habit, more typical of aquamarine.

Figure 51 (above). *Spodumene, variety kunzite, 4.5 cm tall, Beebe Hole, Jacumba district, San Diego County. Recovered in 1974 by Loren Beebe. Fisher-Kureczka specimen.*

Figure 52 (below). *Beryl, variety morganite, 2.5 cm tall, with muscovite and microcline, Hercules dike, Little Three mine, Ramona district, San Diego County. Recovered in 1983 by Lou Spaulding Jr. Fisher-Kureczka specimen.*

Figure 53 (opposite). *Beryl, variety aquamarine, 4.5 cm tall, on microcline, Pack Rat mine, Jacumba district, San Diego County. Cal Graeber specimen.*

Work at the Hercules continued into the early 1990s, when it was curtailed due to safety concerns and difficulty in maintaining an ever-deepening excavation. The claim has since lapsed, and little if any further work has been done there. The Little Three property is still owned by Spaulding. Future prospects for the mine are complicated by encroaching human development nearby.

Jacumba District

Pegmatites have been known on Tule Mountain, near the town of Jacumba, just north of the Mexican border, since the early twentieth century, but only recently has anything of note been produced from the district. In 1974 local prospector Loren Beebe discovered a breached pocket of kunzite exposed on the surface, and he reportedly collected 24 kilograms of spodumene of various colors. Beebe filed a claim on the property, naming it the **Beebe Hole,** but he appears to have found little else there. One of the best specimens from the find, a 22-cm-long bicolored (lilac to pale yellow) crystal, is now in the collection of the LACMNH.

In 1981 a claim known as the **Pack Rat mine** was established adjacent to the Beebe Hole, and the two claims were subsequently merged. During the 1980s and 1990s the Pack Rat produced small amounts of spessartine, aquamarine, columbite, apatite, and a host of rare phosphates. Although several additional pegmatites are known to exist nearby, the BLM has been reluctant to allow development due to the "ecologically sensitive" nature of the area.

Conclusions

During the past century the pegmatites of Southern California have produced numerous world-class specimens of colored tourmaline, beryl (principally morganite), kunzite, and many other pegmatite minerals, which now grace collections, both public and private, around the world. Mining in the region has been cyclic, reaching high points during the first decade of the twentieth century and again during the 1970s and 1980s. For a combination of reasons—including changes in land-use policies, encroaching human development in formerly rural mining areas, and the increasing costs associated with mining—activity is currently at a low point once again. Despite this, finds continue to be made on a few properties that remain active, and in the past ten years notable finds have included those at the Cryo-Genie and Oceanview mines.

ACKNOWLEDGMENTS

Numerous individuals and organizations assisted in the preparation of this article. First and foremost, I would like to acknowledge the past assistance of the late Dr. Eugene Foord, who provided many rare and unpublished documents and photos, including those by W. T. Schaller, as well as much feedback and general inspiration in the study of Southern California's pegmatites. Access to important collections of regional minerals for photography was given by both Cal Graeber and Bryant Harris. Additional photographs were provided by Brendan Laurs, Jeff Scovil, and Mark Mauthner. Photos of mining activities taken by the late Josie Scripps during the early 1970s were generously provided by Cal Graeber. Map images were prepared by William Besse. Dr. Anthony Kampf, of the Los Angeles County Museum of Natural History, and Mark Mauthner provided a useful critique of the manuscript in its various forms. Proofreading of the manuscript and aesthetic consultation regarding the final selection of photos were provided by my wife, Joan Kureczka.

BIBLIOGRAPHY

Černý, P. 1991. Rare-element pegmatites (parts 1, 2). *Geoscience Canada* 18 (2): 49–81.

Fisher, J. 1999. A recent find of kunzite at the historic Katerina mine, Pala district, San Diego County, California. *Mineral News* 15 (1): 1, 6–7.

———. 2000. Les pegmatites gemmiferes du sud de la Californie. *Le Règne Mineral* 31:9–36 [in French].

———. 2002. Gem and rare-element pegmatites of Southern California. *Mineralogical Record* 33 (5): 361–407.

Fisher, J., E. E. Foord, and G. A. Bricker. 1998. The geology, mineralogy, and history of the Himalaya mine, Mesa Grande, San Diego County, California. *Rocks & Minerals* 73 (3): 156–80.

Foord, E. E. 1976. Mineralogy and petrogenesis of layered pegmatite-aplite dikes in the Mesa Grande district, San Diego County, California. Unpublished PhD thesis, Stanford University, Stanford, California.

———. 1977. The Himalaya dike system, Mesa Grande district, San Diego County, California. *Mineralogical Record* 8 (6): 461–74.

Foord, E. E., D. London, A. R. Kampf, J. E. Shigley, and L. W. Snee. 1991. *Gem-bearing pegmatites of San Diego County, California*. Geological Society of America 1991 Annual Meeting field trip guide 9.

Foord, E. E., L. B. Spaulding, R. A. Mason, and R. F. Martin. 1989. Mineralogy and paragenesis of the Little Three mine pegmatites, Ramona district, San Diego County, California. *Mineralogical Record* 20 (2): 101–27.

Geffner, P., and J. Fisher. 1997. Recent work at the Lithia Dike claim, Cahuilla Mountain, Riverside County, California. *Rocks & Minerals* 72 (3): 156–61.

Hanley, J. B. 1951. *Economic geology of the Rincon pegmatites, San Diego County, California.* California Division of Mines special report 7B.

Jahns, R. H., and L. A. Wright. 1951. *Gem and lithium-bearing pegmatites of the Pala district, San Diego County, California.* California Division of mines special report 7A.

Kampf, A. R., K. Gochenour, and J. Clanin. 2003. Tourmaline discovery at the Cryo-Genie mine, San Diego County, California. *Rocks & Minerals* 78 (3): 157–68.

Kunz, G. F. 1905. *Gems, jeweler's materials, and ornamental stones of California.* California State Mining Bureau bulletin 37.

Larsen, E. S., Jr. 1948. *Batholith and associated rocks of Corona, Elsinore, and San Luis Rey quadrangles, Southern California.* Geological Society of America memoir 29.

Larson, W. 1972. The Queen reigns again. *Lapidary Journal* 26:1002–32.

———. 1977. The best of San Diego County. *Mineralogical Record* 8 (6): 507–15.

Mauthner, M. 2008. The 49er Pocket, Oceanview mine, Pala district, San Diego County, California. *Rocks & Minerals* 83 (4): 292–97.

Rynerson, F. J. 1967. *Exploring and mining for gems and gold in the West.* Healdsburg, CA: Naturegraph.

Sinkankas, J. 1959. *Gemstones of North America*, vol. 1. New York: Van Nostrand Reinhold.

———. 1976. *Gemstones of North America*, vol. 2. New York: Van Nostrand Reinhold.

———. 1997. Gemstones of North America, vol. 3. Tucson, AZ: *Geoscience Press.*

Snee, L. W., and E. E. Foord. 1991. $^{40}Ar/^{39}Ar$ thermochronology of granitic pegmatites and host rocks, San Diego County, California (abs.). In *Geological Society of America Annual Meeting abstracts with programs, San Diego, Calif., Oct. 21–24, 1991*, 189.

Stern, L. A., G. E. Brown Jr., D. K. Bird, R. H. Jahns, E. E. Foord, J. E. Shigley, and L. B. Spaulding Jr. 1986. Mineralogy and geochemical evolution of the Little Three pegmatite-aplite layered intrusive, Ramona, California. *American Mineralogist* 71 (3,4): 406–27.

Swoboda, E. R. 2002. History of the Tourmaline Queen mine. *Mineralogical Record* 33 (5): 409–25.

Szenics, T. 1970. The special magic of pink tourmaline. *Lapidary Journal* 24:1042–50.

Todd, V. R., and S. E. Shaw. 1985. S-type granitoids and an I-S line in the Peninsular Ranges Batholith, southern California. *Geology* 13:231–33.

Weber, F. H., Jr. 1963. *Geology and mineral resources of San Diego County, California.* California Division of Mines and Geology county report 3.

TOURMALINE.

Rubellite. Indicolite. Achroite. Aphrizite.

H.=7–7.5. G.=3.0–3.2. Brittle. Luster vitreous. Color black, brown, blue, green, red, colorless. Some specimens are red internally and green externally; others red at one end and green, blue, or black at the other.

The red or pink transparent varieties are called *Rubellite;* if violet-red, *Siberite; Indicolite* is blue or bluish black; *Brazilian Sapphire* is Berlin blue; *Brazilian Emerald, Chrysolite of Brazil,* green and transparent; *Peridot of Ceylon* is honey-yellow. *Achroite* is the name given to colorless tourmalines. *Aphrizite* is black, with a resinous fracture. *Dravite* is brown, greenish black.

A complex silicate of alumina, boron, magnesia, iron, and alkalies (soda, potash, lithia), with small amounts of water and fluorine.

Among the most interesting and beautiful of gem-minerals are the highly colored varieties of tourmaline. It is only recently that they have come to be much known or used in jewelry, though pink (shan) tourmaline has long been greatly and almost superstitiously prized in China. The ordinary tourmalines are black or brown, but some varieties are pink, red, green, and dark blue, and these when transparent make elegant gems. Rarely, they are quite colorless (achroite). Tourmalines are remarkable also for certain optical properties which render them incapable of being successfully imitated, and for the fact that the same crystal will often show two or more richly contrasting colors in different parts.

The first recognition of these minerals in the State apparently goes back as far as 1872, when Mr. Henry Hamilton, in June of that year, obtained some very fine and handsome colored tourmalines on the southeast slope of Thomas Mountain, in Riverside County.

The first discovery in San Diego County is thought to have been made about twenty-five years ago, when some Indian children, at play in a camp near what is now Mesa Grande postoffice, picked up an oddly shaped stone, six-sided like a quartz crystal, about three inches long and a little thicker than a common lead-pencil. On cleaning it off and rubbing it with a bit of hide, it was seen to be of a beautiful blue color, bright and partially clear, almost like a sapphire. The natives had no idea of its nature, but were attracted by its beauty and singularity. Subsequently, other highly colored stones of like character—some blue, others green, others red—were picked up in the same vicinity by Indians and cowboys, but no one realized that they had any actual value.

In Pala, San Diego County, Mr. William Irelan, Jr., State Mineralogist, reported that fine transparent crystals of rubellite (red tourmaline), though not of gem quality, had been found.

The first important development at Pala was announced by Mr. C. R. Orcutt, in 1890.** Here a ledge of lepidolite (lithia mica) containing rubellite was traced for quite a distance. The rubellite crystals are clustered in radiating groups in the fine compact mica; they are not large and not clear, and hence are not suitable for cutting; but their color is a rich rose-red, and they make elegant specimens, on the background of lilac lepidolite. These have gone into collections and museums all over the world, and the material has been mined by tons, partly for specimens and partly for extracting lithia compounds from the lepidolite. More recently, amblygonite (alumina-lithia phosphate) has been found at this mine in large quantities, and this is now the greatest lithia mine in the world.

The rubellite crystals found here are entirely embedded in lepidolite, and until recently it was found impossible to remove them to show their complete form. They were, however, often polished with the lepidolite—the rubellite appearing as pink radiations in a darker gangue of lilac-colored lepidolite. Recently, however, the crystals of rubellite have been worked out, as it were—made to stand out by removing the lepidolite matrix by means of brushes and cleaning-tools—forming most beautiful groups of crystals.

---

* U. S. Census Rept. 1900, Precious Stones, by G. F. Kunz, p. 1050.

** Report on the minerals of the Colorado Desert; 10th Ann. Rept. State Mineralogist of California, 1890. Min. Res. U. S. (Rept. U. S. Geol. Survey). 1893, pp. 17, 18 (reprint).

Reference: adapted from *Kunz, George, 1905, Gems, Jewelers' Materials and Ornamental Stones of California, California State Mining Bureau Bulletin No. 37, June 1905.*

North Adit, Elizabeth R Mine.

Structural Controls and Mineralogical Indicators for the Formation and Distribution of Pockets in the Elizabeth R Mine, Pala, San Diego County, California

Charles E. Houser SCS Engineers
San Diego, California
chouser@scsengineers.com

Based on a senior thesis presented as partial fulfillment of the requirements for Bachelor of Science degree San Diego State University, May 1986.

ABSTRACT

The Pala district in northwestern San Diego County produces fine mineral specimen and semi-precious gemstone material. Part of the Peninsular Ranges batholith, the Pala pegmatites have been mined to various degrees since around the beginning of the twentieth century. Tourmaline, beryl, and spodumene are among the minerals produced. The Elizabeth R mine, located on Chief Mountain in the Ocean View pegmatite, was begun in 1974.

Fine-grained rocks generally occur in the footwall of the dike and may or may not exhibit layers rich in garnet or tourmaline. Graphic granite occurs as masses up to several feet across in all parts of the dike, but primarily in the upper portion near or along the hanging wall. Where no core zone is present, the rest of the dike consists of tourmaline-bearing and muscovite-bearing pegmatite made up of quartz, microcline and albite. Core zones occur in the central or lower portions of the dike and contain the largest crystals in the pegmatite. Primary core zone minerals are quartz, perthite, and spodumene. Pockets or cavities within the core zone contain euhedral crystals of quartz, perthitic microcline, albite, muscovite, tourmaline, beryl, columbite, apatite, and a variety of other phosphate minerals.

The Elizabeth R mine exhibits several controls and indicators for the formation and distribution of significant pockets. Pockets appear where a dike bulges in thickness. These bulges may occur where some weakness, such as a crosscutting fracture set, is encountered in the host rock. The dike may also bulge where the dip changes. These flexures of the dike are referred to as "rolls." In areas where the pegmatite dike does bulge, whether due to either of the structural controls described above or due to other factors, it appears the thicker sections of the dike may facilitate development of a well-developed "core" zone. The core zone is an area of the dike, generally in the central portion, where the largest crystals and the most economically significant mineral assemblages occur.

Certain mineralogic aspects of the pegmatite serve to indicate the presence of pocket-bearing pegmatite. A change in feldspar type, namely an increase in percentage of microcline to albite, occurs in the core zone in and around pockets. Also a change in the chemical composition of large, tapering tourmalines, consisting of a downward enrichment of lithium, is seen near pocket-bearing parts of the dike. Thus changes in mineralogy, either as distinct mineral species or chemical changes within a specific crystal, may indicate the nearby presence of pockets of economically important minerals.

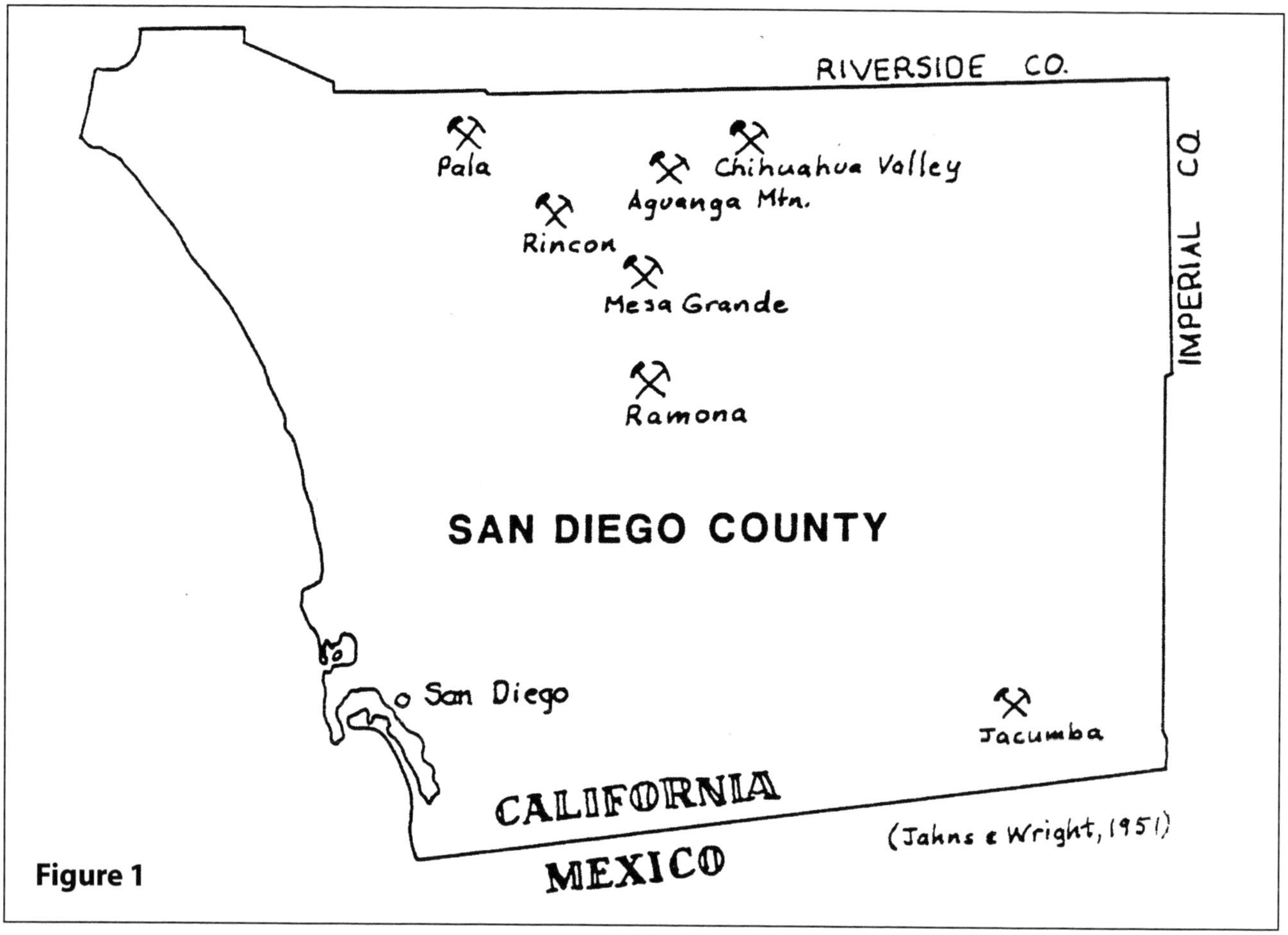

Figure 1

INTRODUCTION

This paper was originally an undergraduate thesis which the author completed in 1986 while working toward his bachelor of science degree in geology at San Diego State University. Although the mine itself has progressed significantly since this study was done, the content of this research and findings of this thesis remain valid.

The pegmatites of Pala, California, are a late-stage part of the Peninsular Ranges Batholith. The Pala district is in northwestern San Diego County about 45 miles north of San Diego (Figure 1). The main part of the district occupies an area of about 5 square miles (Jahns and Wright 1951). Mining began in Pala about the start of the twentieth century and has continued sporadically ever since. Though tourmaline has been the most sought after mineral in the district, several other semi-precious gemstone minerals, including beryl and spodumene, have also been recovered. These various minerals are used both as mineral specimens and as cutting material suitable for faceting into gemstones. Many mines in Pala, including the Stewart, Tourmaline Queen, and White Queen, have been highly acclaimed for their production of high quality material. However, these minerals are not found distributed throughout the pegmatites. Rather they are concentrated in cavities or "pockets" that occur in certain zones in the pegmatite. Because these pocket-bearing zones are not found throughout the dikes, methods have been sought by which to understand the distribution of pockets and predict their occurrence. This report was undertaken in an attempt to determine some of the controls on and indicators for the existence of pockets.

The Elizabeth R mine, known today as the Oceanview mine, is located on the northwest slope of Chief Mountain in the central part of the Pala district (Figure 2). It was chosen as the subject of this report for two main reasons. First is its accessibility to the author and second is the condition of the underground workings. Many mines in Pala and indeed throughout San Diego County are virtually inaccessible to the general public due to the nature of the material being found and the desire of the miners to protect their mines from potential theft. Mines not being worked are generally in poor condition. The Elizabeth R mine is an active mine and had, at the beginning of this study, approximately 575 feet of underground workings. At the completion of this study, this has advanced to about 725 feet. The tunnels are in very good condition and offer good exposures of zonation in the dike.

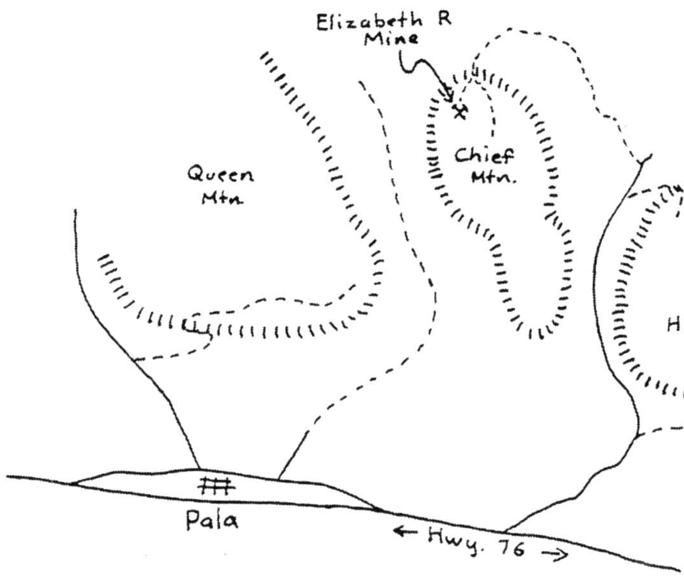

Figure 2 PALA DISTRICT

Two aspects of pegmatites, structure and mineralogy, have been investigated in this study as possible guides to the occurrence and distribution of pockets. Structure includes zones where the pegmatite has intruded weak or fractured host rock, and rolls or flexures in the dike. In sections of the pegmatite exhibiting these types of structure the dike tends to be thicker. It is in these thickened areas or bulges that pockets are generally found. Certain characteristics of minerals in the pegmatite can indicate the presence of pockets. Changes in mineralogy, whether a change in species within a given mineral group or a change in chemical composition across a single mineral grain, are seen in and around pocket-bearing parts of the dike.

MINE HISTORY

The Elizabeth R mine is in what is referred to as the Ocean View pegmatite (Jahns and Wright 1951). The mine is owned and operated by Jeff Swanger following his purchase in 2000 from Roland Reed of El Cajon, California. The only evidence of work prior to 1974 is an open cut approximately 50 feet north of the north adit. Jahns and Wright (1951) report that early production was mainly quartz and beryl. Roland Reed acquired the property from Joe Zeld and in 1974 made the initial cut in a bulge approximately 50 feet up the hill from the area that now contains the north adit. Upon reaching the core, several small pockets were encountered which contained morganite and colored tourmaline as well as quartz, cleavelandite, and apatite (Reed 1983). What is now the north adit was driven to the core and the main drift advanced south. Several large pockets were found containing quartz crystals in groups up to 150 pounds and 30 inches in diameter (Reed 1983). Upon encountering a concentration of lithium phosphates, Mr. Reed began to stope updip and found a series of pockets containing morganite as well as the already noted minerals.

The main drift was then driven south along the strike of the dike. A second adit was driven to the surface approximately 140 feet south of the north adit and so will henceforth be referred to as the south adit. Because this adit seemed to be on the edge of a very pronounced roll or bulge, Mr. Reed decided to stope updip along the south side of a section of the quartz-spodumene core encountered earlier. Several small pockets containing quartz and feldspar were found. Then, 40 feet updip, a large pocket containing a number of superb beryl specimens, including concentrically zoned and bi-color crystals of morganite and aquamarine, was uncovered.

After several short drifts were made off this stope, a new stope was advanced updip just north of the last stope and angled slightly to the north. One rather significant pocket was found about the time the author began spending time at the mine. This pocket contained a few morganites and a large number of milky to clear quartz crystals and crystal clusters.

Upon completion of a short stope mid-way between the adits in which the dike appeared to "pinch out" about 25 feet above the main drift, a new adit was driven into the hillside to intercept the dike approximately 80 feet below the existing workings. Work also began to advance the main drift north from the north adit. At the time of this study in the mid-1980s, mining was continuing at the north end of the main drift and would soon commence in the lower adit (Figure 3). Further discussion of progress and expansion of the mine since completion of this study can be found at the end of this paper.

Figure 3

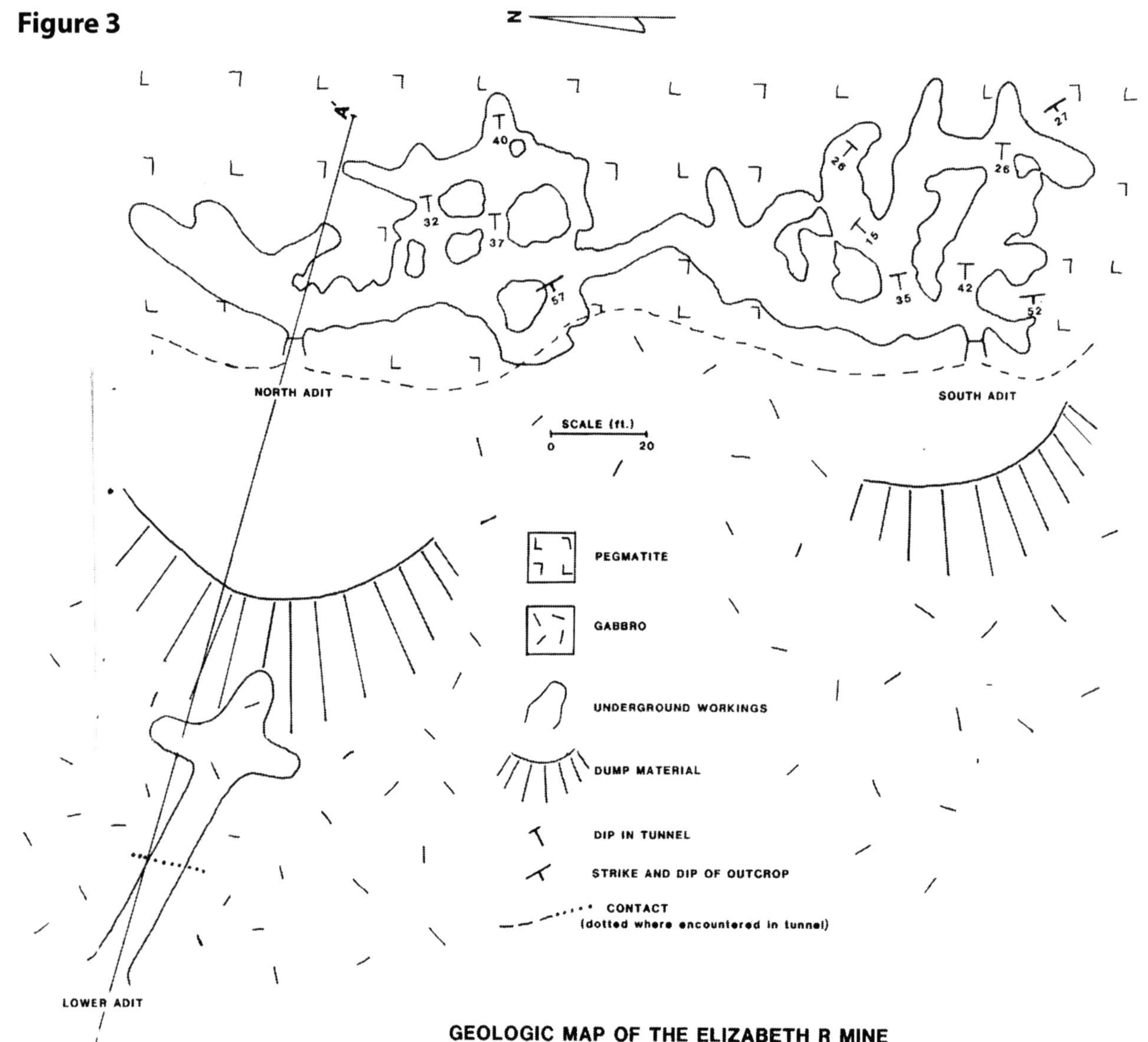

GEOLOGIC MAP OF THE ELIZABETH R MINE
PALA, CALIFORNIA

PETROGRAPHY

Fine-Grained Rocks. The fine-grained rock, normally referred to as line rock, usually occurs in the footwall portion of the dike (Figure 4). It is similar in composition to the rest of the pegmatite, being primarily quartz, orthoclase and albite though the average grain size is significantly smaller, being approximately 1 to 2 millimeters. This unit is referred to as line rock because of the layered appearance of some of these rocks. This is due to alternating thin layers that are rich in garnet or tourmaline. One of the best exposures of line rock in the mine is just inside the south adit in the column to the left of the junction of the main drift and the first stope.

Graphic Granite. One of the more abundant rock types in the mine, the graphic granite forms masses that range from a few inches to several feet and more. It consists of perthitic microcline (Jahns and Wright 1951) containing grains of quartz. The quartz grains occur as elongate rods that vary in cross sectional shape. Some have a triangular or sub-rounded section whereas others have a platy or L-shaped section (Jahns and Wright 1951). The quartz rods range in length from less than an inch to a foot or more and in width from 1 1/16 inch to as much as 1 inch (Jahns and Wright 1951).

The graphic granite can occur in all parts of the dike but is normally found in the upper part of the dike near or along the hanging wall (Jahns and Wright 1951) (Figure 4). In some parts of the mine the graphic granite comprises a substantial part of the pegmatite. In the southern section of the upper workings, visual inspection of the outcrop and the underground workings gives the impression that the upper one-third to one-half of the dike is composed largely of graphic granite. This is difficult to ascertain as no location in the mine contains a complete section of the dike. However, this abundance of graphic granite would be expected as graphic granite forms the most widespread rock type in the Pala pegmatites (Jahns and Wright 1951). Contacts between the graphic granite and surrounding rock types in the dike are commonly gradational and so not very clear.

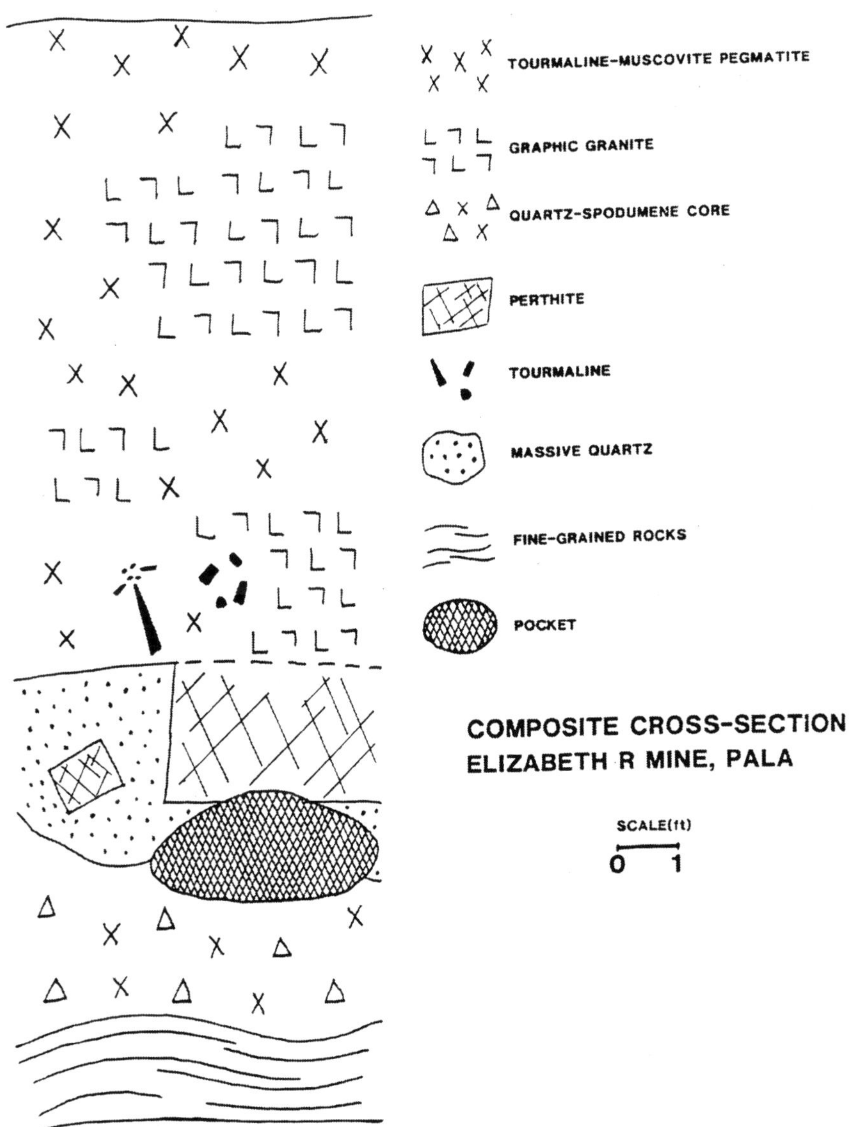

Figure 4

One more noticeable association is the gradation of graphic granite into subhedral or euhedral perthite. This is seen in both pocket-bearing and non-pocket-bearing parts of the dike.

Tourmaline-Muscovite Pegmatite. This rock type occurs alone as a unit and dispersed between and commonly through the blocks of graphic granite. Its primary constituents are intergrown anhedral crystals of quartz, microcline and albite with sporadic muscovite and tourmaline. Muscovite has two modes of occurrence, as individual books and foils ½ inch to 2 inches in diameter and as flakes 1⁄16 inch to approximately ½ inch across disseminated through the rock. The areas or "pods" of disseminated muscovite vary from a few inches to 8 feet across.

Tourmaline occurs as individual euhedral to subhedral crystals ranging from a fraction of an inch to 12 inches or more in length. A pattern is sometimes seen in the orientation of the tourmaline crystals. They are found oriented normal to the hanging wall of the dike in a comb-like structure. In addition, they may form a radial group of crystals radiating out from a finer grained core of tourmaline. Most commonly the tourmaline crystals are randomly distributed through the rock. In the Elizabeth R mine, this rock unit is found in all parts of the dike. In sections without a core zone it commonly makes up a major portion of the dike. The dike in such sections may be composed entirely of graphic granite and tourmaline-muscovite pegmatite with roughly equal distribution of both. In some sections of tunnel, the entire exposed section appears to be tourmaline-muscovite pegmatite. Near the south adit this rock type is observed below the line rock that is normally seen at the bottom of a pegmatite dike.

Core Zone. Jahns and Wright (1951) use the term "giant pegmatite" to describe the core zone. This is a fitting term as it is the core that contains the largest crystals found in the pegmatite dike. Though pockets are found in all parts of the dike, a vast majority of them occur in the core zone. Hence it is the unit of greatest economic importance. The presence or absence of core zone material in an outcrop is one of the best indicators whether a pegmatite has the potential to be productive. The Elizabeth R mine exhibits very well developed core zones that range in thickness from just a few inches to as much as 6 or 7 feet. The core zones are tabular in cross section and taper toward the edges. In plan view (Figure 5) these zones vary in shape depending on the nature of the controls that influence their development. Core zones are found in the thickest parts of the dike and the importance of this observation will be discussed in detail in later sections of this report.

Core zones in the Elizabeth R mine generally occur in the lower portion of the dike (Figure 4), though this is difficult to determine without a

Figure 5

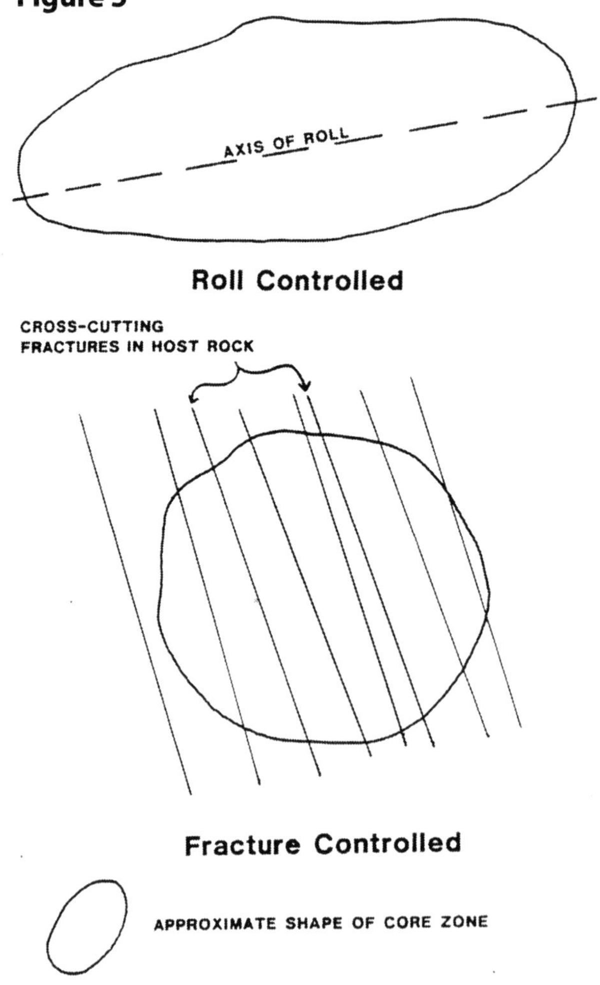

STRUCTURE-DEPENDANT SHAPE OF CORE ZONE

completely exposed section of the dike. The core in the southern part of the upper workings rests directly on the line rock (Figure 6). This relationship is clearly visible just inside the south adit at the junction of the main drift and the first stope. In the northern part of the mine, the core is not nearly as well defined as in the southern part. This may be due to the size of the zone as it appears to be somewhat larger in the north area. Recent work (1986) north of the north adit reveals a large core zone up to 6 feet thick showing an excellent potential for pockets.

The most abundant mineral in the core zone is perthite. It occurs as subhedral to euhedral crystals 1 inch to 6 feet across. As mentioned earlier, these crystals commonly grade upward out of the core zone into graphic granite. The most common color of perthite is white to cream. It sometimes occurs in a gray-brown color. This gray-brown color, coupled with the well defined cleavage of the perthite, gives it a distinctly different appearance from the other minerals common in the core zone. The perthite also occurs as euhedral crystals enclosed in massive quartz. A lattice-like appearance commonly characterizes the perthite. This is because of preferential leaching of either the albite or the microcline from the perthite. The perthite is commonly altered and may be the source of some of the clays found in the pegmatite.

Quartz is another major constituent of the core zone. It occurs as subhedral to anhedral crystals and as massive pods that vary in size up to 5 feet or more in maximum dimension. The intergrowth of euhedral perthite in massive quartz commonly gives a rather striking appearance. Quartz is considered by the miner to be a good indication that pockets may occur in the core although it seems an over-abundance of quartz can indicate the opposite.

Commonly associated with the quartz are lath-like crystals of highly altered spodumene. These range from an inch in maximum dimension to as much as 15 inches by 2 inches. This association can be seen in the column supporting the winze in the northern section of the upper workings and also occurs extensively in the southern section. One stope in the southern section exhibits a pod of phosphate minerals 2 feet across containing an impressive intergrowth of the altered spodumene crystals (Figure 7). Locally a core of unaltered spodumene will be present in the center of the altered spodumene crystal, but these are very small and economically insignificant.

Figures 6 and 7.

One very interesting rock that occurs in and adjacent the core zone is a type of feldspar referred to by miners as "de-silicified feldspar." This is somewhat a misnomer since feldspar itself is a silicate mineral. The rock gets its name because of the apparent selective leaching of the quartz rods from the feldspar matrix. This gives the rock a vuggy, honeycomb-like appearance. These pods occur adjacent pocket bearing parts of the dike and were probably graphic granite before alteration occurred (Jahns and Wright 1951). Tourmaline is commonly present in these rocks with crystals that average about two centimeters (Simpson 1965).

Pockets. Pockets occur as cavities in the pegmatite as small as ½ inch and as large as 10 feet long by 5 feet wide by 4 feet high. Figure 8 shows quartz crystals in a clay-filled pocket. The minerals lining the pockets become euhedral as they grow into the pocket and pockets larger than a few inches are filled with crystals of various minerals. Clay minerals commonly fill the pockets and give the rock surrounding the pockets a reddish brown color due to the iron staining from the clay. Pockets nearly always occur in the core zone of the dike but all parts of the dike have been observed to contain pockets.

By far the most common mineral found in pockets is quartz. It is normally milky but also occurs as beautiful clear and smoky crystals. These crystals range from a fraction of an inch in length to 18 inches and possibly more.

Perthitic microcline and orthoclase are also common. Orthoclase is white and microcline is white to tan in color. The microcline is commonly altered by the removal of the albite to form the common latticelike appearance. Some of these crystals will also have an overgrowth of unaltered orthoclase. Microcline and orthoclase crystals are normally 1 to 2 inches in maximum dimension although much larger crystals have been noted.

Albite occurs in several forms in the pockets. One form is the fine-grained "sugary" albite that occurs not only in the pockets but throughout the core. Two habits of cleavelandite variety of albite are found in pockets in the Elizabeth R mine. One is the bladed variety found in mines throughout San Diego County. The blades rarely exceed ½ inch in length and form attractive matrices for quartz and beryl crystals. Bladed cleavelandite is very light tan to white, and rarely clear. Another form of this mineral is the blocky crystals that range up to 1 inch across. These are white to light tan and are quite a common constituent of pockets in parts of the pegmatite.

Muscovite occurs in many of the pockets. It is silvery white to green in color and is most commonly associated with bladed cleavelandite and quartz, though it is also found associated with beryl, blocky cleavelandite and tourmaline.

Beryl occurs in roughly half of the major pockets. Most common in the Elizabeth R mine is the peach colored variety called morganite. The blue-green variety, aquamarine, is found in the core of some morganites. It is very rarely found comprising a whole beryl crystal. Beryl crystals range from less than an inch to 6 inches across.

Tourmaline is also found in the Elizabeth R mine, but it is largely limited to a certain area of the mine. This zone of tourmaline pockets is along the south edge of the southern section of the upper workings and forms a line of pockets running roughly parallel to the dip of the dike. The pocket tourmaline crystals are elbaite with a very dark blue color: however crystals larger than 1/16 of an inch in diameter appear black. Tourmaline crystals occur in a variety of sizes, the largest being a crystal measuring 7 inches long by 2 inches wide. Another fine tourmaline specimen consists of a tourmaline crystal perched atop a large quartz crystal. The specimen is 10 inches high and the tourmaline is 1½ inches across. Minerals which occur much less frequently in pockets in the Elizabeth R mine include columbite, apatite and a variety of other phosphate minerals.

It should be noted that the ability to efficiently locate pockets is the motivation behind this entire study as it is the pockets that contain virtually all the economic mineral deposits in the pegmatite. One large specimen of beryl crystals on a matrix of quartz and cleavelandite (Figure 9) was valued by Mr. Reed at $20,000 to $30,000 in the mid-1980s. One cannot underestimate the importance of finding pockets in mining a pegmatite. It is with this in mind that this investigation has been made into determining controls and indicators for the existence of pockets.

Figure 8

Figure 9

a.

b.

c.

d.

e.

f.

g.

From left to right, top to bottom:

a. Quartz crystal cluster (approximately 8 inches high)

b. Goethite pseudomorph after pyrite? (Large pseudomorph approximately ¾-inch across)

c. Tourmaline crystal (approximately 1¾ inches high)

d. Spodumene, var. kunzite (lilac colored) with light green "cap" (approximately 2½ inches long)

e. Columbite crystal (approximately 1 inch across) on muscovite and cleavelandite

f. Tourmaline partially replaced by lepidolite. (Crystal is approximately 2½ inches across)

g. Morganite beryl and quartz (specimen approximately 6 inches long)

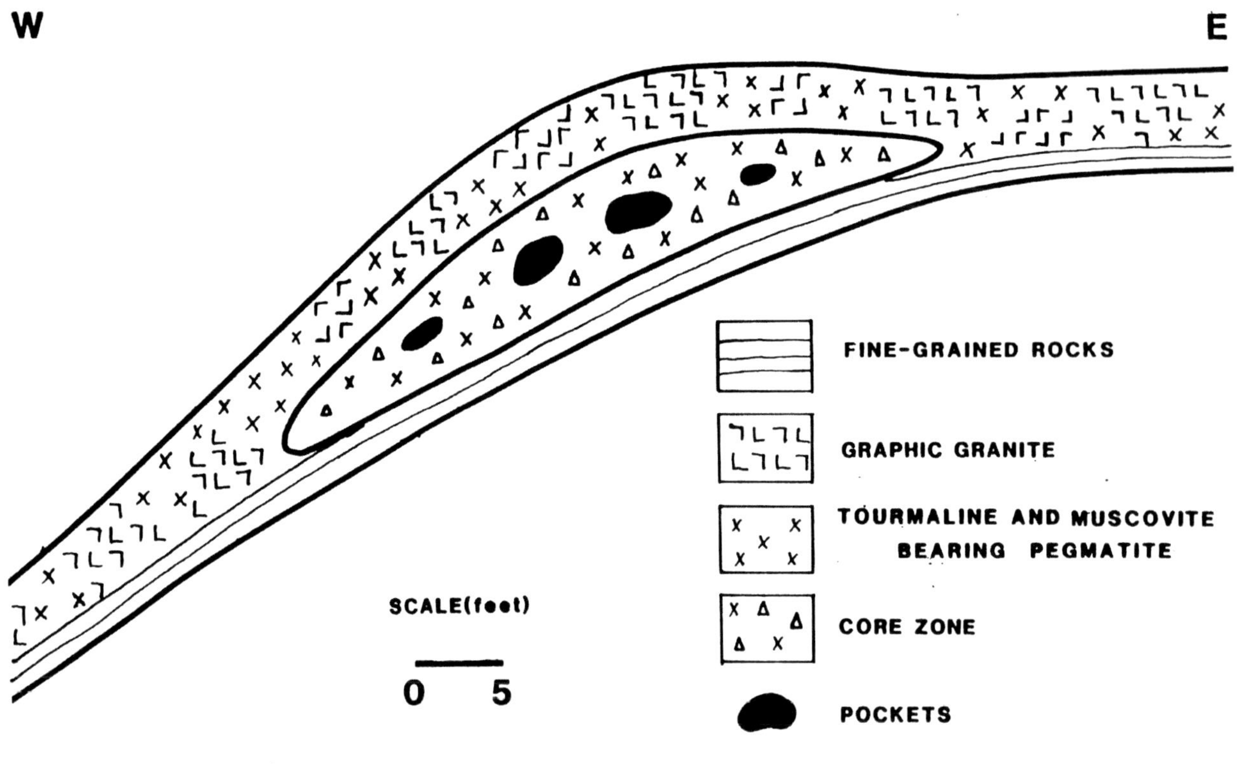

Figure 10 **Roll Controlled Bulging**

STRUCTURAL CONTROL

Two aspects of structure were investigated as possible controlling factors for the formation and distribution of significant pockets. The first, which required little more than visual inspection of the outcrops and underground workings, is that of flexures or "roll" in the pegmatite dike. The second is the major fractures in the pegmatite, presumably an expression of fracture systems in the host gabbro. This feature required more detailed field measurements and correlation with regional features using aerial photographs. The following discussion will explain in more detail the influence of structure on the formation and distribution of pockets and observations that indicate the presence of these structures in the Elizabeth R mine.

Various individuals have stated that rolls in pegmatite dikes are necessary for the formation of pockets. The distribution of pockets along a roll is explained by an analogy that compares the relationship between rolls and pockets to oil being trapped in the apex of an anticline as the oil rises through the higher density host rock. This analogy may seem reasonable in a pegmatite system as one would expect the volatiles to rise. If this were true, pockets would tend to occur along antiformal rolls with synformal rolls being virtually barren of any economic or exotic mineral deposits. Such is not the case and two observations serve to rule out this explanation. First, pockets are commonly distributed along synformal rolls as well as antiformal rolls. Therefore, even if the entire system has been overturned since emplacement, the pockets are still most commonly evenly dispersed between synformal and antiformal rolls. The second observation is that certain pegmatite dikes, such as the Himalaya dike in Mesa Grande, have no noticeable roll structure, dip angles in excess of 40 degrees, and still contain significant distributions of highly mineralized pockets (John Mclean 1986, personal communication). An alternate explanation, which is considered the direct effect of the two structural controls discussed in this paper, is that pockets form where the pegmatite dike is allowed to bulge in thickness and thus develop a well zoned structure including a very-coarse grained core zone. It is in these bulges that the volatile phase is allowed to develop, the coarse core being a reflection of this fact. If the volatile phase develops to a sufficient extent, the volatiles can collect to form large pockets. Therefore, any factor allowing the pegmatite to bulge could contribute to the development of pockets.

It has been noted that where a pegmatite rolls or flexes, it tends to be thicker. This may be due to some inherent weakness in the host rock that allows the dike to bend and at the same time expand as it intrudes (Figure 10). Several observations serve to indicate a roll structure in the Elizabeth R pegmatite dike. The shape of the outcrop, if it is assumed that the face of the outcrop approximates the hanging wall of the dike, indicates a significant roll or thickening in the southern section of the upper workings. Immediately above the mine dump, the outcrop dips 52 to 57 degrees, whereas above this it flattens out to only 27 degrees. The shape of the underground workings, namely the stopes that are driven updip from the main drift, also reflect this roll. The dip of the stopes below the roll dip up to 42 degrees with dip angles as low as 15 degrees above the roll (Figure 3). Again an assumption must be made – that the core zone, which determines the shape and geometry of the tunnel as it is the core that contains the significant pockets, will reflect the overall structure of the dike. A number of pockets have been encountered in the vicinity of this apparent roll, including two pockets that contained several hundred pounds of material each (Figure 11). One of these pockets contained a single specimen of quartz and feldspar (now in the author's back yard) with an estimated weight of 350 pounds and requiring four people to lift!

Another observation is made on the relative attitude of the dike in the upper workings compared to the attitude in the lower adit. Pegmatites tend to vary in dip to form terrace-like features (Jahns and Wright 1951). As was mentioned, the dip of the dike in the lower part of the upper workings is relatively steep, up to 42 degrees. In the lower adit, 140 feet down the hill, the dike dips 27 degrees where the tunnel first encounters the dike. It also contains fractures, interpreted to be cooling fractures, that

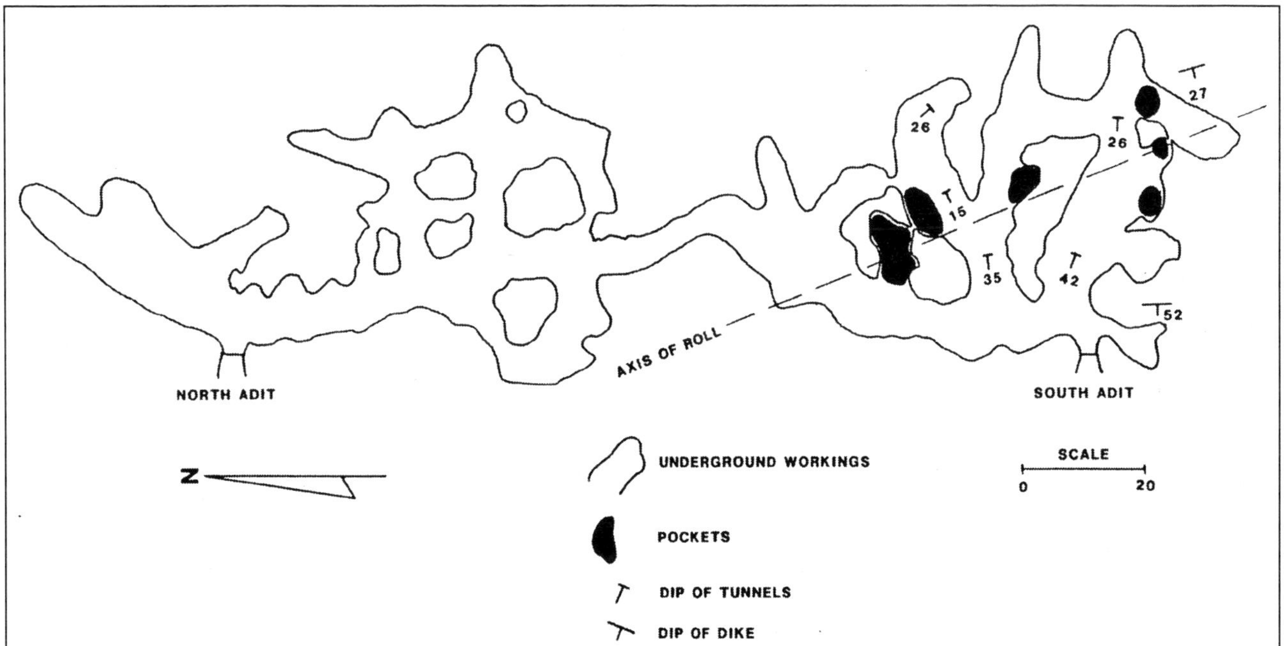

Figure 11 ROLL STRUCTURE-SOUTH SECTION OF UPPER WORKINGS ELIZABETH R MINE, PALA

Figure 12

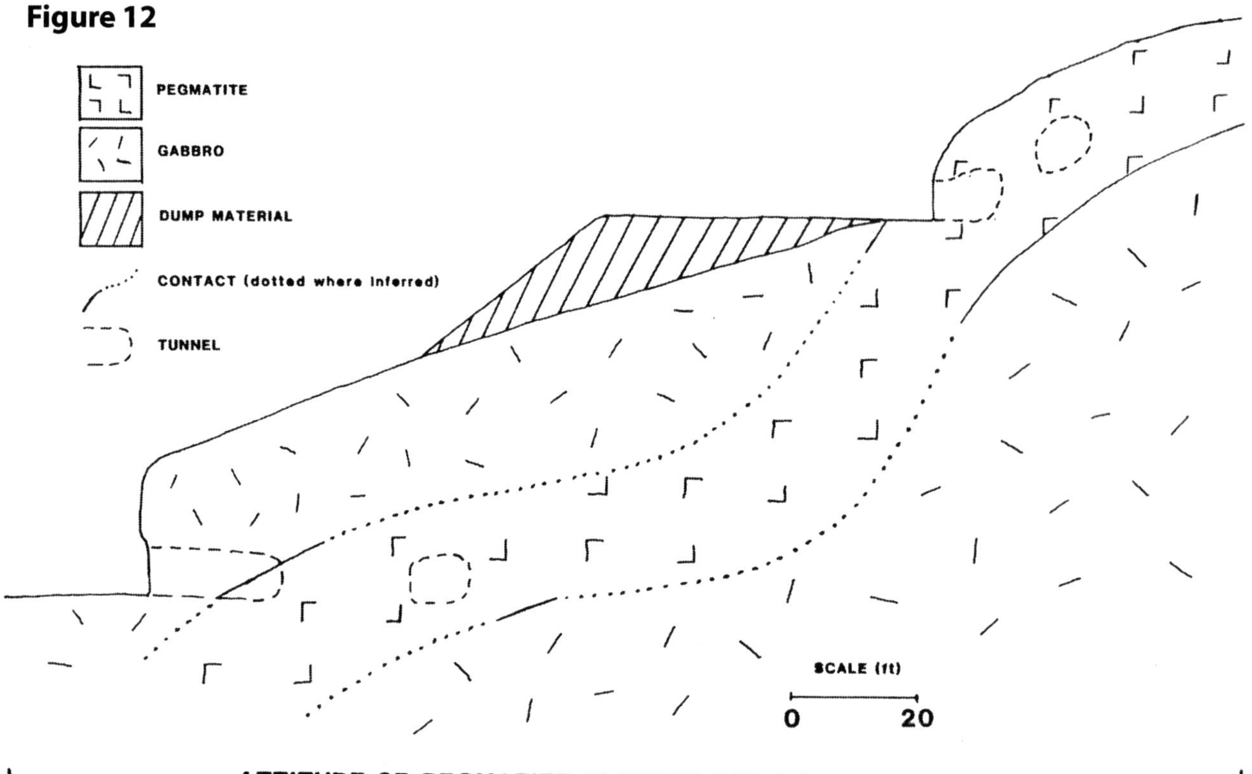

ATTITUDE OF PEGMATITE IN UPPER AND LOWER WORKINGS
ELIZABETH R MINE, PALA

are parallel to the contacts of the dike, and that dip only 18 degrees near the end of the tunnel. Extrapolation of the dike from the lower adit to the upper workings indicates an upwards roll, potentially rather abrupt, between the upper and lower workings (Figure 12). This roll is of particular interest as no exploration has yet reached this section of the dike. The potential for significant pockets to be encountered in this section of the pegmatite warrants some excitement and anticipation on the part of those individuals involved with working the mine.

As was discussed earlier, the dike can bulge where it encounters weak or fractured host rock. Attitude measurements were taken on major measurable fractures throughout the upper workings. Thirty-one measurements were made with 16 measurements (52%) appearing to fit into a sub-parallel set that strikes N60E to N80E and dips 60 to 85 degrees north (Figure 13a). These fractures occur in the northern and southern sections of the upper workings with a zone approximately 50 feet wide in the center nearly void of any major fractures (Figure 13b). There is a contact between the gabbro and another igneous pluton, presumably tonalite or granodiorite, approximately ¼ mile north of the mine. This contact is, at the point closest to the mine, roughly parallel to the fractures observed in the mine workings. The rocks north of the contact show a distinct lineation which is also parallel to the contact and the mine fractures. These lineaments strike N73E to N76E north of the mine (Figure 14). The gabbro, however, shows only very weak lineation at one, and possibly two, locations near the mine. This is attributed to the nature of gabbro as it tends to develop a very deep and very strong weathering profile that would weaken or obliterate any display of lineation. Though it is very weak, the lineation observed in the gabbro also parallels the mine fracture set, the contact, and the lineation in the rocks north of the contact. Therefore the fracture set in the mine appears to have some correlation to the larger scale structural features of the area.

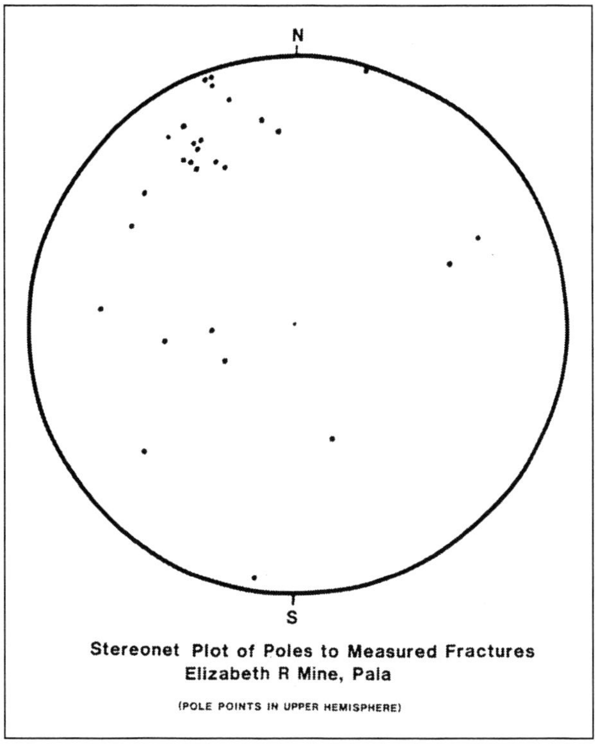

Figure 13a

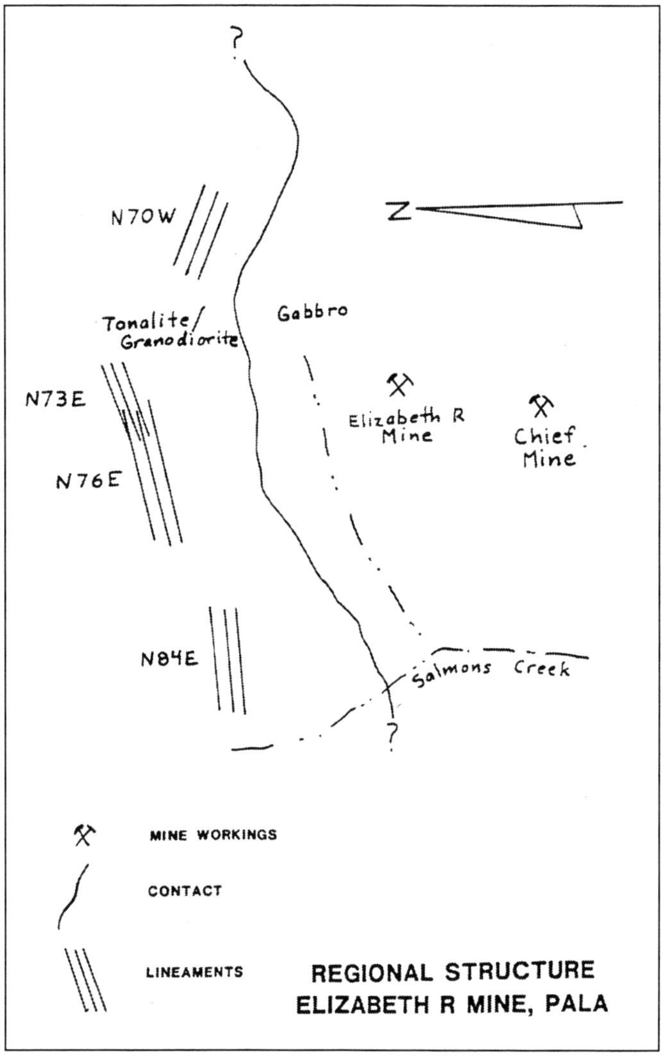

Figure 14

Figure 13b

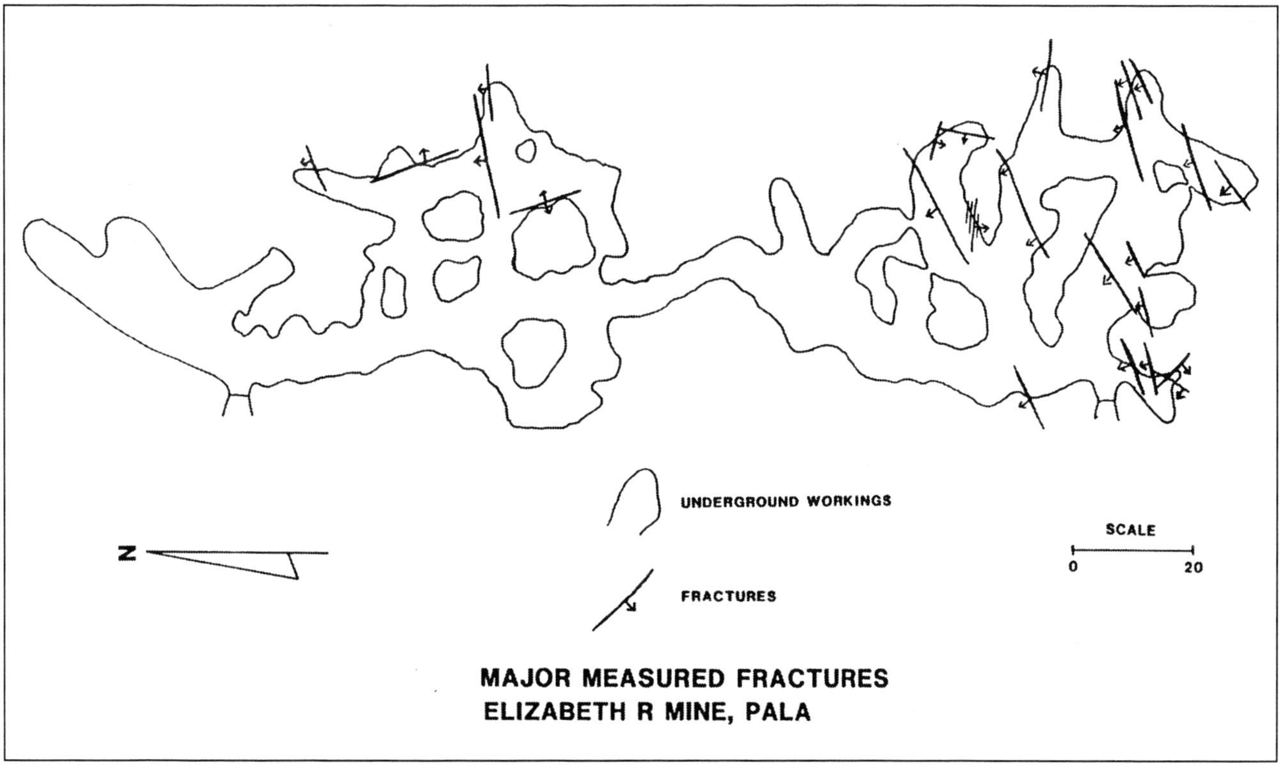

Joints in the gabbro are not necessarily spaced evenly. They may occur in zones, giving the rock a sheet-like appearance, with areas between free of joints for up to 50 feet (Jahns and Wright 1951). These joint sets or zones would represent a weakness in the rock. If, as the dike intrudes, it encounters such a zone, it might bulge and thus fulfill the criteria for the formation of one or a series of pockets. Though no clear jointing is ascertainable in the host gabbro, the fractures in the northern and southern part of the upper workings may be an expression of pre-existing fracture zones in the gabbro (Figure 15). The unfractured and barren area in the center of the workings would then be in a relatively un-jointed section of the gabbro.

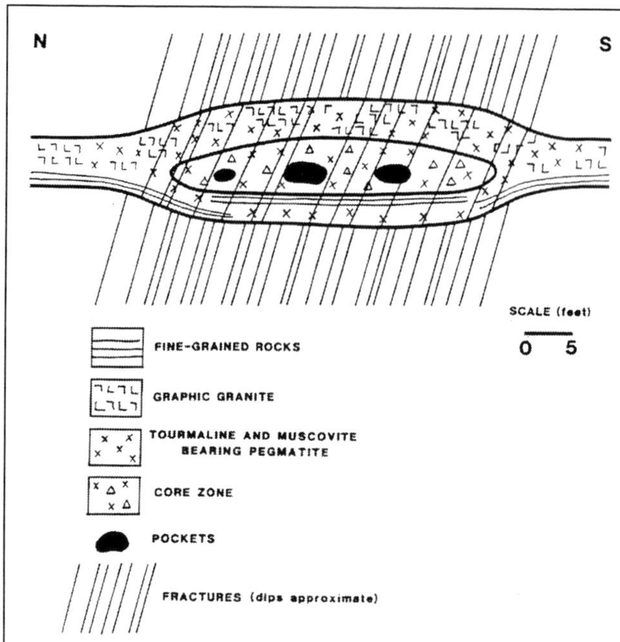

Fracture Controlled Bulging **Figure 15**

MINERALOGICAL INDICATORS

Two mineralogical investigations were undertaken to determine whether or not any changes in mineral occurrence or chemistry might indicate the presence of potential pocket bearing pegmatite. Samples of feldspar were collected from various positions in and above the core zone and x-ray diffraction used to find what, if any, differences occur. Tourmalines were also tested by x-ray diffraction to determine any difference between the top and bottom ends of large, tapering crystals that occur just above the core zone. The idea was to see how their chemistries might change and, if possible, in what elements they become enriched as they approach pockets or pocket-bearing dike.

Feldspars. Microcline and albite are the two primary feldspars present in the core zone of the pegmatite. In a sample taken above the core zone, microcline is represented by ten major peaks whereas albite is only represented by one major peak. A sample from near the top of the zone has microcline represented by seven peaks and albite by one. A mid-zone run has eleven microcline peaks and two albite peaks. A feldspar sample taken from just below the core zone shows a distribution of nine microcline peaks and eight albite peaks. Converting this to a percentage of major peaks, the albite occupies 9.1% above the core zone, 12.5% in the top of the core zone, 15.4% in the middle core zone and 47.1% below the core zone. This does not necessarily reflect the exact percentage of microcline to albite, but indicates relative amounts of each mineral. The ratio of microcline to albite increases substantially upward through the core zone, indicating a general upward increase in potassium.

Tourmaline. Some larger tourmalines above the core zone in the Elizabeth R mine have been observed to taper upward. Another way of looking at this, obviously, is that they widen as they grow downward. This is thought to be due to increasing volatile content as the dike crystallizes inward allowing the mineral grains to grow larger. Because of this widening of the crystals, it was decided to test the top and bottom of several individual crystals to determine if any chemical difference occurred as the crystal grew downward. Two crystals were tested. One crystal is located above the pocket on the north side of the northernmost stope of the southern part of the upper workings (Figure 16a). The other tourmaline is across from the first one above the core zone where the stope splits (Figure 16b).

Figures 16a and 16b.

The x-ray diffraction run shows the top of the first crystal to be schorl. With the bottom of the crystal four major peaks are schorl and two are elbaite. The same is true with the second crystal, but with two peaks schorl and three elbaite in the bottom. This shows an increase in lithium content as the crystals progress downward.

The occurrence of either of these two mineralogical changes, a change in the type of feldspar or a shift in composition of the tourmaline, might indicate the nearby existence of pocket bearing pegmatite. However, their usefulness is questionable since in the time spent preparing the samples and running the tests one could easily advance workings enough to determine whether core zones or pockets are an imminent discovery.

MINING UPDATE

Since completion of my undergraduate thesis, the Elizabeth R mine has undergone extensive expansion, and even experienced a name change. Mr. Reed continued mining, primarily in the lower workings (beginning with the "lower adit" depicted on Figure 3), adding a south adit to the lower workings, connecting the north and south lower adits in similar fashion to the upper workings, and stoping updip to intercept the upper workings in several locations. He also explored the pegmatite in a downdip direction. Production in the upper workings included quartz crystals, schorl tourmaline, and morganite (sometimes with "aquamarine" cores in the crystals), as well as attractive green, elongate muscovite crystals on or off matrix. With advancement of the tunnels in the lower workings since the completion of this thesis, not only have more pockets of fine morganite and quartz specimens been recovered, but aquamarine beryl crystals have been found and a number of pockets, particularly toward the southern end of the lower workings, have yielded colored tourmaline and kunzite!

Mr. Reed has "semi" retired from mining, although he still makes visits to the workings to view the progress and get a "fix" of the adventure. In 2000, Mr. Reed sold the mine to Mr. Jeff Swanger, who changed the name to the Oceanview mine, after the name given to the pegmatite dike early in the history of mining in Pala. Mr. Swanger has continued mining, adding additional workings even further down-dip below Mr. Reed's lower workings, and continues to recover morganite, colored tourmaline, and very high quality kunzite specimens, as well as fine aquamarine beryl crystals on cleavelandite matrix. He has also built a business providing visitors not only opportunities to see a working pegmatite mine, but to actually "mine" the deposit by screening material brought out of the mine during on-going operations. A visitor typically finds specimens of beautiful pegmatite minerals such as quartz, lepidolite, and the various gem minerals, and if lucky enough, may find a truly high quality tourmaline, beryl, or kunzite that simply got away from the miners. The 2013 SDAG field trip has the privilege of visiting this class operation to screen for the beautiful minerals and see the underground workings of this world class pegmatite mine in the famous Pala mining district.

ACKNOWLEDGMENTS

The first and greatest thanks goes to Roland Reed for opening the mine to this investigation and supplying numerous references and information for its compilation. The late Dr. Michael Walawender of San Diego State University agreed to serve as advisor for this thesis and originally suggested the subject of this report. Jim Clanin, formerly of Pala and now living in Maine, originally introduced the author to Mr. Reed and contributed immeasurably to the author's initial knowledge of pegmatite systems. Others who helped with field mapping, photography, and thoughtful discussion during the "working it out" phase of this project include David Swanson, Paul Sherman, and Larry Neuvirth. Mr. Swanson also supplied the computer and the necessary instruction in the use of word processors.

I wrote this thesis on a Commodore 64. It took five minutes just to load the word processor (from a 5½ inch floppy disc) before the thesis file (stored on a 5½ inch floppy disc) could be uploaded and written or edited. There was as yet no such thing as a mouse, except of course the small, gray, furry creatures which were abundant at that time.

REFERENCES

Jahns, R.H., and Wright, L.A. 1951. *Gem and lithium-bearing pegmatites of the Pala district, San Diego County, California.* Special Report 7-A. State of California Department of Natural Resources.

Simpson, D.R. 1965. *Geology of the central part of the Ramona pegmatite district, San Diego County, California.* Special Report 86. Short Contributions to California Geology. California Division of Mines and Geology.

Reed, R. 1983. Morganite from the Elizabeth "R" Mine, Pala, California. unpublished report.

"You know, I used to like this hobby. ... But shoot! Seems like *everybody*'s got a rock collection."

The hills are shadows, and they flow
From form to form, and nothing stands;
They melt like mist, the solid lands,
Like clouds they shape themselves and go.

— Alfred Lord Tennyson

The Tin Can Flat Landslide of Palomar Mountain, San Diego County, California

Michael W. Hart *P.O. Box 261227*
Consultant *San Diego, California 92196*
mwhart@aol.com

INTRODUCTION

Many types of previously little-studied landslides including earthflows, wedge-slides, complex slides, and rock avalanches have occurred in the Peninsular Ranges of southern California. This fact is remarkable because, for the most part, the ranges are made up of crystalline rocks of the Peninsular Ranges batholith with pre-batholithic and volcanic rocks underlying a much smaller area. In this study, one of these large landslides is described. Also presented is a hypothesis that explains the probable mechanisms of failure and runout.

The Tin Can Flat landslide, named after a prominent topographic bench located near the head of the landslide, is the largest of many similar landslides that occur on the southern slopes of the Palomar Mountains in San Diego County, California (Figures 1 and 2). These landslides were first recognized by the California Division of Mines and Geology and appeared on a geologic map of San Diego County compiled by Weber (1963). However, no details of the landslides were discussed in the text of his report.

The Palomar Mountains, and the contiguous Agua Tibia Mountains to the northwest, are an approximately 30 km long, rectangular fault block bounded by the Elsinore fault on the south and the Aguanga fault on the north. This block is made up principally of Cretaceous plutonic rocks of the Peninsular Ranges batholith. Pre-batholithic rocks, chiefly Julian Schist, locally form large roof pendants and near the Tin Can Flat landslide make up the principal rock type at the head of the landslide. The lower part of the slide mass consists primarily of weakly-foliated tonalite.

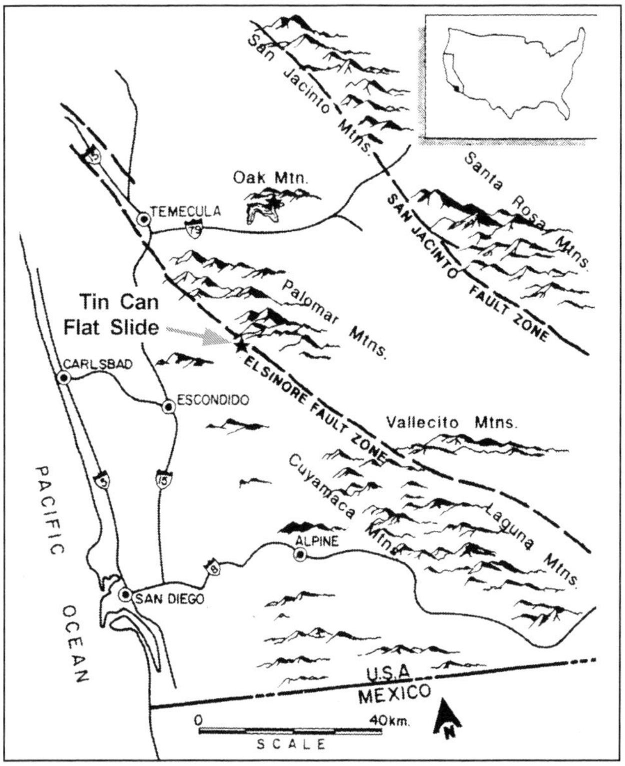

Figure 1. *Location Map*

SLIDE GEOMETRY AND MORPHOLOGY

The Tin Can Flat landslide is approximately 3 km long, 3 km wide, and up to 150 m thick. The southern distal edge of the landslide is partially buried by fanglomerates emanating primarily from Harrison Canyon (Figure 2). The northern part of the distal edge of the slide, however, is well exposed along the banks of Pauma Creek and several of its minor tributaries. Erosion along one of these tributaries has exposed the slip surface and overlying breccia shear zones (Figure 3, Point "1"). The eastern limit of the landslide where it lies in contact with plutonic rocks, is not well exposed. However, aerial photographs that predate the present orchards, show low lateral and distal pressure ridges that represent the limits of landsliding.

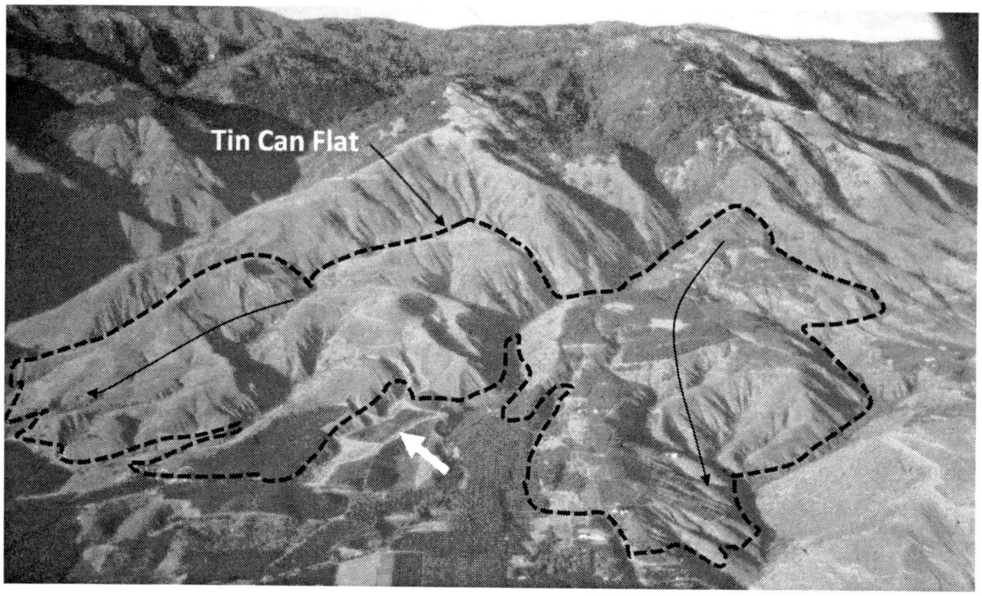

Figure 2. *Tin Can Flat landslide. Pauma Creek at left edge of photo; Jaybird Creek to right of Pauma Creek; Harrison Canyon, center; Sycamore Canyon near right edge of photograph. White arrow: Nate Harrison Grade.*

The arcuate headscarp above Tin Can Flat is well preserved and not heavily modified by erosion. This is in contrast to the highly eroded southern portion of the scarp at the head of Harrison and Sycamore canyons.

The slide surface is characterized by topographic benches that step down toward the southwest, and probably represent the heads of small, secondary landslides. Three major canyons dissect the landslide; Jaybird Creek, Harrison, and Sycamore canyons (Figure 2). Harrison and Sycamore canyons cut completely through the landslide and provide good exposures of its lithology, structure, basal rupture zone, and substrate.

COMPOSITION, STRUCTURE AND CONTACT RELATIONSHIPS

The headscarp above Tin Can Flat is underlain by well-foliated pre-batholithic schist and gneiss. Foliation planes strike nearly parallel to the general trend of the scarp and dip 55° to 85° to the southwest. Gneissic rocks form the highly dissected scarp area south of Tin Can Flat.

Exposures along Nate Harrison Grade and in Harrison and Sycamore canyons reveal the structure and composition of the landslide mass. In the upper part of Harrison Canyon below Tin Can Flat the steep, southwesterly dipping slip surface is exposed near the base of a wide zone of sheared and crushed rock (Figure 3, Point "2"). Relatively unfractured, poorly-foliated, gneissic tonalite occurs below the shear zone and in hand-specimen appears identical in composition to the tonalite located above the shear zone.

Most of the distal slide mass is composed of thoroughly sheared and crushed granitic and metamorphic rocks that display flow banding and slip surfaces within clayey gouge zones. The contact between the fractured but relatively intact-looking rock mass that makes up the eastern part of the slide and the completely comminuted western part is not exposed, but is probably gradational as depicted on Figure 4.

The base of the western portion of the slide is well-exposed in the 15 m to 20 m high, vertical stream cuts of Harrison and Sycamore canyons (Figure 3, Point "3"). Here the slide debris, consisting of thoroughly sheared or comminuted, flow-banded materials with fragments generally less than a few millimeters in size is in gradational contact with an approximately lm thick, moderately well-cemented, crudely-stratified and deformed, dark brown to grey, flow-banded mudstone. The mudstone is, in tum, underlain by fanglomerate consisting of well-rounded, granitic and metamorphic cobbles and boulders in a well-sorted matrix of medium- to coarse-grained sand. Contact relationships between the fanglomerate and slide debris indicate a minimum runout onto a of approximately 1700 m (Figure 4). The contact preexisting

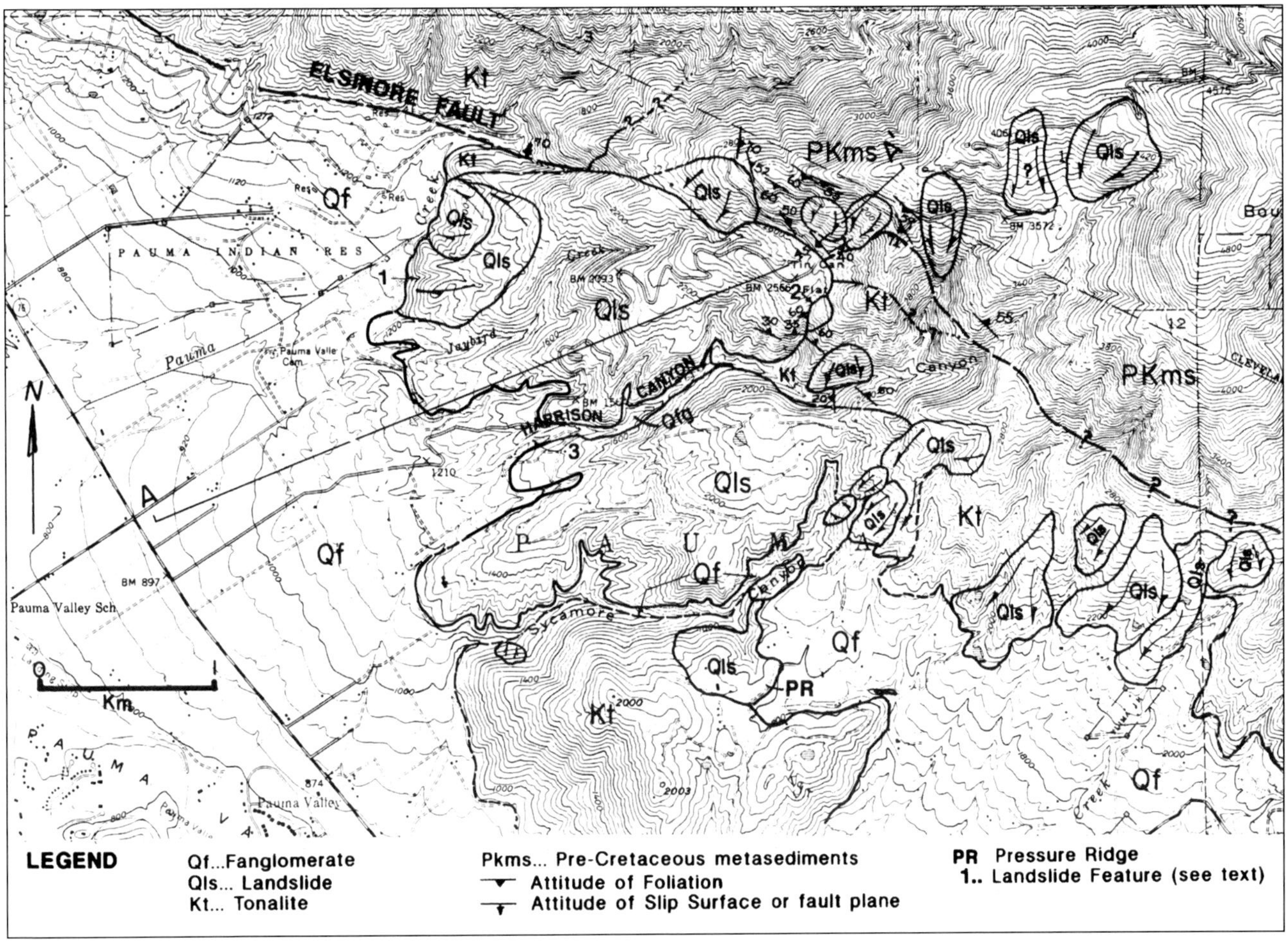

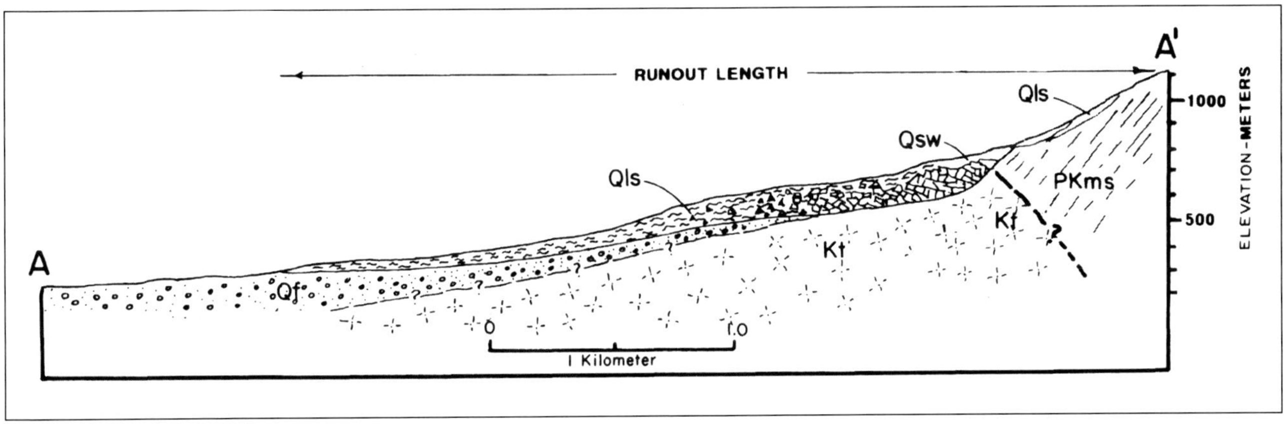

Figure 3 (top). *Geologic map of Tin Can Flat landslide*

Figure 4 (bottom). *Geologic section A-A'*

alluvial fan between the slide breccia and mudstone is sharp but no slip surface is present.

In Harrison Canyon a clastic dike (Figure 5) extends approximately 1 m upward from the mudstone unit into the slide breccia. Similar clastic dikes have been reported in the Blackhawk landslide by Johnson (1978).

MECHANICS OF FAILURE AND RUNOUT

The mode of failure and mechanics of runout are of two of the most interesting aspects of landsliding. To determine the mechanics of failure, one must know the basal rupture surface and slide mass characteristics, geometry, and something of the landslide's erosional history. For example, the contrast between the well preserved headscarp above Tin Can Flat and the nearly obliterated portion of the headscarp to the southeast of

Tin Can Flat suggests that there may have been at least two episodes of sliding separated by a sufficient amount of time to nearly erase the oldest headscarp. However, the stratigraphy of the slide debris on the west side of the canyon is identical to that on the east. Also, the basal rupture zone lies about the same elevation on the west and east sides of the canyon. It, thus, seems more likely that the landslide occurred as a single event, but that various components of the slide have been subject to differing rates of erosion.

The presence of a steep shear zone between fractured blocks of slide debris and in-place gneiss at the head of the slide and the broad topographic bench at Tin Can Flat suggest slumping as a mode of failure. Such failure was, therefore, probably initiated along foliation planes or along a possible continuation of the Elsinore fault that has been mapped to the western edge of the landslide (Vaughan 1987). The presence of flow-banded and comminuted rock in the western three-quarters of the slide and a long runout distance of 1700± m over fanglomerate suggests a rock-avalanche mode of failure. Most rock-avalanches, of which the author is aware, however, are completely detached from their source area and have travelled great distances relative to their drop height over nearly horizontal terrain. The ratio of height drop to length of runout (H/L) is a commonly used measure to determine if a landslide has excessive runout length and whether or not it can be classified as a "classic" rock-avalanche. Rock-avalanches generally have H/L ratios of less than 0.5 (Melosh 1987). The Tin Can Flat landslide has a H/L ratio of approximately 0.24, a ratio that could qualify it as a long runout rock-avalanche.

Figure 5. *Base of Tin Can Flat landslide in Harrison Canyon. A = comminuted landslide debris, B= 1 m long clastic dike; C= Lithified slurry of silt, sand, and cobble-sized clasts making up basal shear zone.*

The above observations suggest a hybrid mechanism of failure; namely, slumping at the head of the slide followed by a rapid runout of the slide debris on a pressurized mud-slurry represented in outcrop today by the flow-banded mudstone in Harrison Canyon. The clastic dike at this location is evidence of the depressurization of the mud-slurry after the avalanche came to rest.

REFERENCES

Hart, M.W. 2003. Landslides in the Peninsular Ranges, southern California, *in* Murbach, M., and Hart, M.W., eds., *Geology of the Elsinore Fault Zone*, Field trip guidebook of San Diego Association of Geologists and South Coast Geological Society. pp. 193-209.

Johnson, B. 1978. Blackhawk landslide, California, U.S.A., *in* Voight, B., ed., *Rockslides and Avalanches, v.1, Natural Phenomena.* Elsevier, Amsterdam. pp. 481-504.

Melosh, H.J. 1987. The mechanics of large rock avalanches, *in* Costa, J.E., and Wieczorek, G.F., eds., *Debris Flows/avalanches; Process, Recognition, and Mitigation, Reviews in Engineering Geology v. VII.* Geological Society of America. pp. 41-49.

Vaughan, P. R. 1987 Alluvial stratigraphy and neotectonics along the Elsinore Fault at Agua Tibia Mountain, California. unpublished M.S. Thesis. San Diego State University.

Weber, H. W. Jr. 1963. *Geology and Mineral Resources of San Diego County, California, County Report 3*, California Division of Mines and Geology, 305 p.

Scientific Truth

Every great scientific truth goes through three stages.
First, people say it conflicts with the Bible.
Next, they say it has been discovered before.
Lastly, they say they always believed it.

—Louis Agassiz

They [plates] can't curl down; they must curl up
To form a kind of dish
To stop the oceans spilling out
And losing all the fish.

— *B.C. & G.C.P. King*

Escondido Canal Landslides and History, San Diego County, California

Gregory T. Farrand
Ninyo & Moore
gfarrand@ninyoandmoore.com

Hans van de Vrugt
Southwest Geophysics
hans@southwestgeophysics.com

GEOLOGY AND LANDSLIDES

In 1993 and 1994, the authors performed a geologic reconnaissance evaluation as part of a condition assessment to identify potential geologic hazards along a 12-mile segment of the Escondido Canal in eastern San Diego County, California. The canal is owned jointly by the City of Escondido and Vista Irrigation District. The canal is located south of the San Luis Rey River (Figure 1), downstream of Henshaw Dam. The canal area studied in 1993-1994 extends southwesterly from the diversion inlet structure in a canyon on the San Luis Rey River (about 8 miles southwest of Henshaw Dam) to the point where the canal discharges into Lake Wohlford. Two penstocks (the Lake Wohlford Penstock extending from Lake Wohlford Dam to the Bear Valley Hydroelectric Plant, and the Rincon Penstock between Horsethief Canyon and Bee Canyon near the Rincon Indian Reservation) were also evaluated. Below the diversion structure in the San Luis River River, the canal is crossed by major drainages such in Bee Canyon, Horsethief Canyon, Lower Hellhole Canyon (Hell Creek), Woods Valley, Escondido Creek in Bear Valley at the upper end of Lake Wohlford, and by several unnamed drainages. Rodriguez Mountain occurs about 1 mile south and 2 miles east of the canal.

A team consisting of the design engineer, City of Escondido staff, Vista Irrigation District staff and Ninyo & Moore's geologist walked the entire canal alignment, noting areas of damage and repairs to the concrete and gunite lined canal, eight tunnels (lined and unlined) up to ½-mile long, seven flume structures, drainage overchutes, an inverted siphon at Hell Creek, and diversion structure with a Parshall flume. We also mapped areas of geologic instability, such as landslides, rock falls, mudflows, slope failures, and precarious boulders impacting the canal.

This paper briefly discusses a mudflow that was mapped along a short segment of canal located east of the Rincon Penstock and west of Bee Canyon (Figure 1 – Inset). A short history of the canal and its construction is discussed based on sparce available documents and historical photos (Photos 1 through 20) listed below. The only dates on the photos are between 1914 and 1924. Historical photos and documents were provided by the City of Escondido Water Department, the Escondido History Center, and other sources.

Bedrock in the vicinity of Escondido Canal consists predominantly of Cretaceous aged granitic rock of the granite of Bottle Peak, granite of Dixon Lake, quartz bearing diorite, granodiorite, and gabbro. The rocks are extensively fractured along portions of the canal.

In the study area for this paper (Figure 1 - Inset), the large mudflow mapped east of the Rincon Penstock and west of Bee Canyon extends from the canal area down to San Luis Rey River, an elevation difference of approximately 700 feet and length of about ½ mile. Extensive smaller slope failures, slumps, mudflows, rockfalls and loose precarious large boulders were mapped along the entire 12-mile long canal, and within the area of the large mudflow west of Bee Canyon. The Bee Canyon and mudflow areas have historically been a chronic maintenance issue for the canal. The granitic rocks upslope and lateral to the mudflow consisted of weathered and fractured granite, granodiorite, and gabbro.

HISTORY OF THE CANAL: THE QUEST FOR WATER

In the late 1800s population growth was on the rise in the Escondido (meaning Hidden) Valley and so was the demand for water. Water wells and local creek water provided early settlers with a sufficient supply of water. However, increased agriculture and drought rendered the existing water system inadequate. The City of Escondido incorporated in 1888 and soon thereafter the Escondido Irrigation District (EID) was formed to provide water to the Valley. The EID sold water bonds and began to break ground on the construction of a new water system. The system included the construction of the Bear Valley Reservoir Dam, later renamed Lake Wohlford Dam after A.W. Wohlford, a prominent leader of water development in the Valley. In addition, a ditch line (canal) roughly 15 miles long would bring much needed water from the San Luis Rey River to Bear Valley. The dam was originally constructed of rock and redwood facing (Photos 17 and 18) and was approximately 76 feet high. The canal consisted of wooden flumes and unlined ditches excavated by hand (Photos 1 and 20) with an occasional dynamite blast. Laborers are believed to have consisted of local inhabitants and possible Chinese workers which were in the area and responsible for constructing many buildings. The capacity of the new water conveyance system was roughly 35 cubic feet per second (cfs).

On July 5, 1895, the gate of the reservoir was opened and the new system was in operation. It wasn't long before problems with the somewhat crude system were realized. Water loss from leaky flumes, gopher holes, and percolation resulted in only about 60 percent of the diverted river water reaching the dam. In 1897 the unlined portions of the canal was either lined with cement plaster or replaced with pipe in an effort to improve deliverance. In addition, repairs to leaky flumes were made.

Problems persisted. A mountain fire in 1904 destroyed a large section of flume. Portions of the conveyance system were deteriorating, and obtaining funds for needed repairs and maintenance was a heated issue. In about 1910 the Escondido Land and Town Company, which owned a large amount of Escondido Mutual Water Company stock (Escondido Mutual Water Company was formed to replace the EID) approved taxes to finance upgrades to the system. Flumes were replaced with concrete-lined canals, and tunnels were added including a large 1,900-foot long tunnel constructed through a portion of Rodriquez Mountain. Near disaster struck, however, when a portion of the redwood faced dam was damaged with dynamite by saboteurs after the United States had entered World War I. Despite the attack, the dam remained structurally intact. In 1924 a new larger dam, approximately 100 feet high, was constructed with an outlet tower structure (Photo 19).

Since the construction of the Lake Wohlford Dam and Escondido Canal, water demands have continued to increase, and fires, mudslides, rock falls, erosion, and material deterioration continue to plague the system. In the last 100 years the system has undergone improvements and repairs to sustain water delivery to the valley. Today the canal consists of six tunnels (one up to ½ mile long), steel flumes, a 2,000-foot long siphon, miles of covered and uncovered open concrete lined channel and reinforced concrete pipe which have doubled its capacity and improved its efficiency.

Recent studies have shown that portions of the existing dam are seismically unsafe and plans for a new dam are underway. The new dam will raise water levels 15 feet and is expected to be completed in 2017 (U-T San Diego 2012).

ACKNOWLEDGEMENTS

We wish to thank Mr. Craig Whittemore and Mr. Carl Burgess with the City of Escondido Water Department for kindly providing most of the historical photos taken during construction of Escondido Canal, and Ms. Wendy Barker with Escondido History Center for assistance in our research of historical documents.

REFERENCES

Buskirk, N, Buskirk, S, Parks, L.M., Hasselberg, J.B., and Langness, P. 1993. *Escondido, Then and Now*. Heritage Publishing Company.

City of Escondido Water Department. 2013. Historical photos of the Escondido Canal taken between 1912 and 1924, courtesy of Mr. Craig Whittemore and Mr. Carl Burgess.

Kennedy, M.P. 1999. Geologic Map of the Valley Center 7.5′ Quadrangle, San Diego County. California Division of Mines and Geology.

McGrew, A.B. 1988. *Hidden Valley Heritage*. Blue-Ribbon Centennial History Committee.

Ninyo & Moore, 1995, Limited Geotechnical Evaluation Escondido Canal and Wohlford Penshock, San Diego County, California, January 12.

Ryan, F.B. and Ryan, L.C. 1980. *Escondido, As it Was 1900-1950*. Taylor Publishing Company.

U-T San Diego. 2012. Lake Wohlford to get a new dam: story by J. Harry Jones, dated February 16.

U.S.Department of Agriculture (USDA).1953. stereoscopic aerial photos.

Weber, H.F. 1963. Geology and Mineral Resources of San Diego County, California, County Report 3. California Division of Mines and Geology.

ILLUSTRATIONS

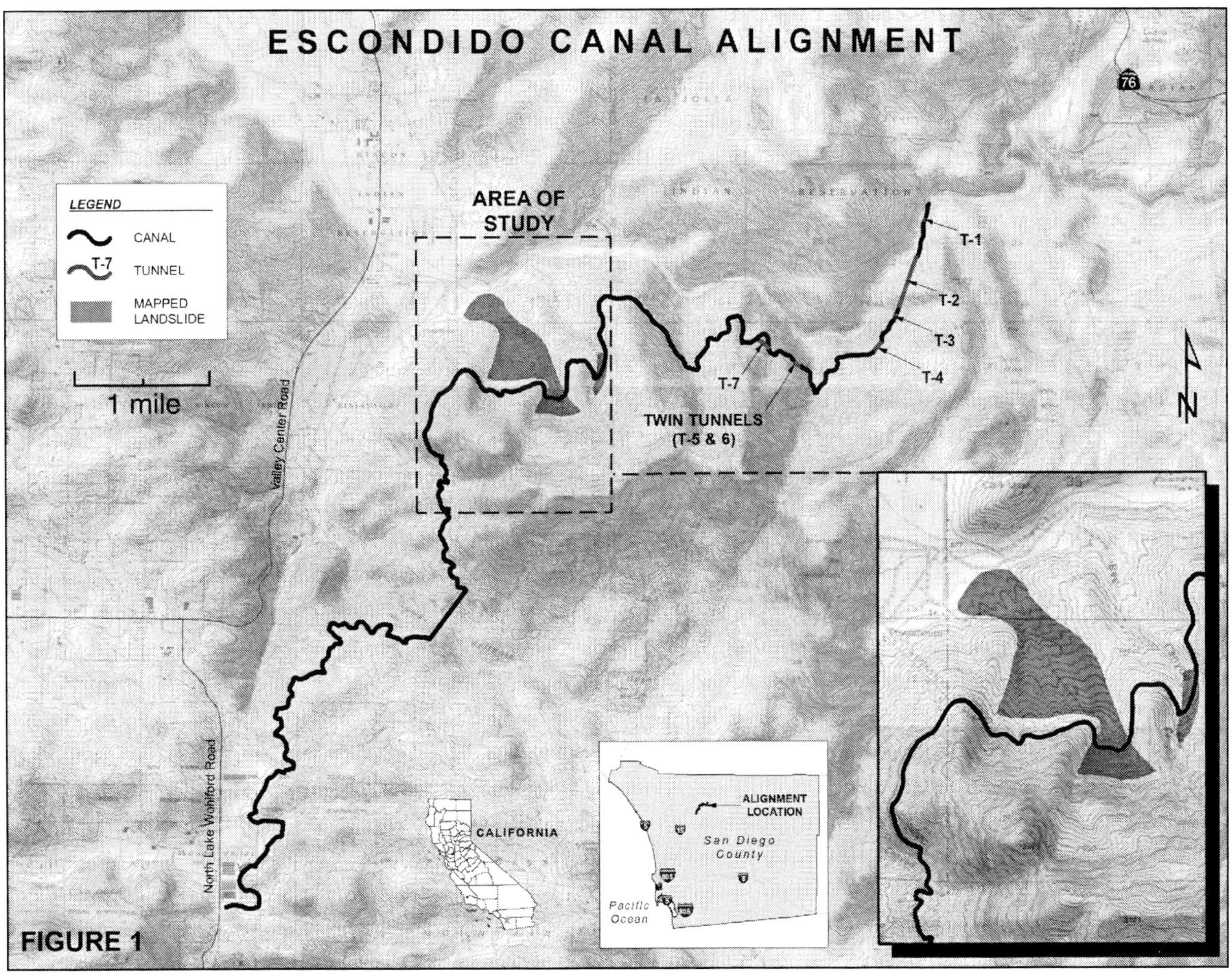

Figure 1. *Escondido Canal alignment, with study area inset*

Photo 1. *Workers in hand excavated canal.*

Photo 2. *Canal with Horsethief Canyon in background.*

Photo 3 (left). *Vertical cliff called Cape Horn by old timers.*

Photo 4 (right). *Cape Horn on trail around Tunnel 2.*

Photo 5. *Outlet of Tunnel 4.*

Photo 6. *Outlet of Tunnel 4.*

Photo 7. *Canal Diversion in San Luis Rey River.*

Photo 8. *Below Diversion in San Luis Rey River.*

Photo 9. *Possibly located near Tunnel 2.*

Photo 10. *Workers inside Tunnel 2; December 1912.*

Photo 11 (above). *View west from flume going around Tunnel 2.*

Photo 12. *Between Rincon Penstock and Horsethief Canyon.*

Photo 13. *Tunnel camp with shacks on canal.*

Photo 14. *Finished canal section.*

Photo 15. *Canal construction in 1924 (5 photos).*
Photo 15a. *Sand bins.*
Photo 15b. *Finished lining.*
Photo 15c. *Enlarging canal.*
Photo 15d. *Sand plant at river.*
Photo 15e. *Placing lining.*

Photo 16. *Canal construction in 1924 (5 photos).*
Photo 16a. *Sand cars, canal railway.*
Photo 16b. *Sand hoist, Rincon Penstock.*
Photo 16c. *Sand truck, Rincon.*
Photo 16d. *Mixing station, canal lining.*
Photo 16e. *General view of main canal.*

Photo 17. *Bear Valley Dam (Lake Wohlford Dam), a rock-filled dam with redwood facing (a popular place for fishing). Source:* Escondido: Then and Now.

Photo 18. *Bear Valley Dam (Lake Wohlford Dam), a rock-filled dam with redwood facing. Source:* Escondido: As it Was.

Photo 19_ *Fifteen miles of canals and flumes were constructed by hand with an occasional dynamite blast. Source:* Escondido: Then and Now.

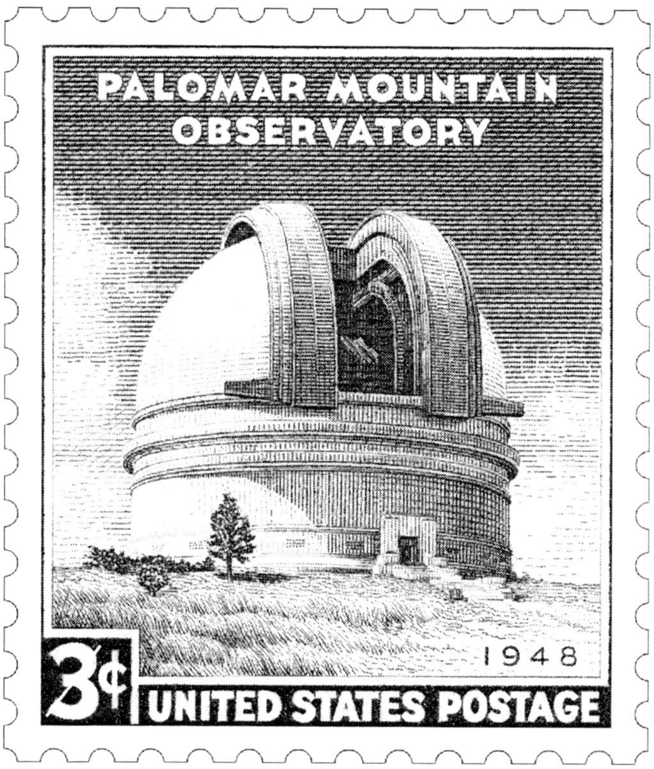
PALOMAR MOUNTAIN
OBSERVATORY
1948
3¢
UNITED STATES POSTAGE

San Luis Rey River – Longitudinal Profile

William J. Elliott Consulting Engineering Geologist
P.O. Box 541
Solana Beach, California 92075

ABSTRACT

The longitudinal profile of the San Luis Rey River includes a prominent knickpoint near Harrah's Rincon Casino. This elevation jump indicates the presence of an immature river grade. Peninsular Ranges uplift and gentle westward tilting, most likely during the recent geologic past, is attested to by an elevation difference of about 800 feet across 2¼ miles (355½ feet per mile gradient). The old graded profile above the knickpoint has, therefore, been separated from the new, geologically young, pro-grading profile between the Pacific Ocean and Harrah's Rincon Casino. With passage of sufficient geologic time, and other things being equal, it is anticipated that the knickpoint will eventually erode its way far enough upstream to once again establish an equilibrated toe-to-head profile more typical of a mature river system.

INTRODUCTION

Located in northern San Diego County, California, the 68½-mile long San Luis Rey River discharges into the Pacific Ocean near the northwest corner of Oceanside, California, at Mile Zero (SW¼, Section 22, T11S, R5W, SBBM; Figure 1). Its headwaters are in a saddle along the crest of the Bucksnort Mountains (NW¼, NW¼, NW¼, Section 7, T9S, R4E, SBBM). There are several alternate choices for the headwaters of the San Luis Rey River along the rim of *Valle de San Jose* [Portilla] (also variously known as Warner Valley, Warner Basin, and Henshaw Valley and Henshaw Basin). The channel chosen follows the *named* San Luis Rey River beyond Lake Henshaw, to and past Indian Flats Campground, past Chihuahua Valley, and to the headwaters of Johnson Canyon (Figures 1 and 2).

Profile elevations and distances were taken from United States Geological Survey 7½′ topographic quadrangle map sheets (scale 1′=2,000′, contour intervals: 20 feet or 40 feet, with some intermediate contours at 10 feet). See references. Where a precise stream thalweg was not indicated, the center of the river bed was chosen as an estimated line of profile. Distances are thought to be accurate to within approximately 0.025 mile. Elevations are believed to be accurate to within one half of a contour interval.

The San Luis Rey River watershed is home to three airports, Oceanside, Pala (private), and Warner Springs (glider-port). Four Native American gambling casinos can be found within the San Luis Rey River drainage basin: Pala, Pauma, Rincon, and Santa Ysabel. It is also home to the San Luis Rey and Pala missions, as well as several recreation/resort venues and golf courses. Historically, mining of many flavors, as well as agriculture and ranching have been hosted by the San Luis Rey River Watershed.

The river is herein arbitrarily divided into six segments, starting with Segment 1 at the Pacific Ocean.

SEGMENT 1

From the river's mouth at the Pacific Ocean (Photo 1), Mile Zero, to near the "T" intersection of North River Road and Hermosa Drive (between Vandegrift Boulevard and Vista Way [at River-Mile 10.18, approximate river elevation 100′]) the nearly flat river valley varies in width from as narrow as about 500 feet to locally wider than about 5,000 feet. Photo 2 shows the Old River Road bridge (SR-76) at River-Mile 11.93. Average gradient is

9.8 feet per mile. Except for some mapped areas of granitic rocks east of Vandegrift Boulevard, the river bed is bordered principally by Eocene and younger clastic sedimentary deposits (Weber, 1963, Pl. 1; Rogers, 1966; Weber, 1982, Pl. 1).

Over the years, ground water for domestic and agricultural use has been pumped from numerous wells in the relatively shallow sandy aquifer. Additionally, construction aggregate (mostly sand) was mined from the river bed prior to about 1990, when the City of Oceanside began to aggressively enforce the State of California Surface Mining and Reclamation Act of 1975 (SMARA). Currently, the lower reaches of the river are largely channelized, with the adjacent flood plain being checkered with residential and commercial urbanization.

Lawrence Canyon Aggregate Mine.* The mouth of Lawrence Canyon is located immediately south of State Route 76, about 0.4-mile east of Interstate 5. Aggregate (sand and gravel) for use in local construction projects was mined on this 100 acres site during the 1950s from lower and middle Miocene San Onofre Breccia (Elliott and Marshall, 2012, p. 119; Weber, 1963, p. 248, Pl. 1, Sequential Catalog No. 403).

Mining. See Weber (1963, p. 201, 235, Pl. 1) and accompanying text for descriptions of past sand, gravel, decomposed granite, and peat mining in this segment of the San Luis Rey River watershed.

SEGMENT 2

The next stretch of river, upstream from about the "T" intersection of North River Road and Hermosa Drive to about River-Mile 38.98, has an average gradient of about 29.86 feet per mile. Over a distance of approximately 28.8 miles, the river bed rises to an elevation of approximately 960 feet. The upper point of this segment is located in a narrowing stretch of river canyon (S½, Section 31, T10S, R1E, SBBM) about 2 miles east of Valley Center Road bridge — not far from Harrah's Rincon Casino. Throughout this lengthy stretch, the river valley continues to be relatively flat and wide (again, from as narrow as about 500 feet to locally wider than about 5,000 feet).

Most of the river's banks are bordered by mid-late-Cretaceous crystalline granitic and older metamorphic rocks (Weber, 1963, Pl. 1; Rogers, 1966).

Historical river-bed uses have included: water extraction, sand mining, resort and recreation, rural residential, orchards, farming, ranching, and dairy. Some of these activities and uses continue today.

San Luis River Flood Plain, Segments 1 and 2. For a good visual of the length and *width* of the San Luis Rey River flood plain, see the eye popping contrast between the long wide yellow line of river alluvium and adjacent red granitics on Ellis and Lee's Plate 3 (1919).

Rosemary's Mountain Quarry. Located on the north side of State Route 76, approximately 1⅓ miles east of Interstate 15, this open-pit quarry supplies the construction industry with a wide variety of crushed rock and rip rap. A lengthy permitting process, which began in 1987, was completed in 2004. On-going commercial operations began in 2008 (Perazzo and Perazzo, 2012). Larsen (1948, Pl. 1) identified the quarry rock as Indian Mountain leuco-granodiorite.

Proposed Gregory Canyon Landfill. The proposed Gregory Canyon landfill site is located on the south side of the San Luis Rey River (south of State Route 76), approximately 3⅓ miles (by road) east of Interstate 15. Proposed more than 20 years ago, this approximately 300 acre site received voter approval for use as a landfill in 1994 and 2004. Situated on the west side of Gregory Mountain (an unnamed 1844' peak [U.S.G.S., 1968c]), the proposed solid waste facility is to be constructed in a small, short, unnamed valley that discharges seasonal rainfall into San Luis Rey River. Ongoing obstacles to construction include: fear of possible ground-water and San Luis Rey River pollution, as well as Native American concerns. Viewed from State Route 76

* **Indented Paragraphs:** *Geologic highlights and points of historic interest associated with each segment are noted by indented paragraphs.*

(Photo 3), the Gregory Mountain skyline profile is easily identified as an asymmetrical hill mantled with scattered large white weathering boulders of Indian Mountain leuco-granodiorite (Larsen, 1948). This bouldery outcrop pattern is sometimes loosely referred to as *woolsack* topography (for example, Larsen, 1948, p. 77).

Pala Mission. This *assistencia* (branch of Mission San Luis Rey), was founded in 1815. Located on the north side of State Route 76, it is the only California mission with a detached bell tower (Jahns, 1954a, p. 28–29, Map 8, Figures 9 and 10; Figure 3, Photos 4, 5 and 6). A stroll through the mission chapel (Photo 7) and around its grounds provides a sense of the life and times of California's early 19th century Mission Period.

In recent years, this once small sleepy backcountry community has become a popular Native American gaming destination.

Pala Tourmaline Mines. North and east of the community of Pala, world famous tourmaline mines can be seen from State Route 76. Notable mine claims include: Stuart (Photo 8), Pala Queen, Pala King, Chief and Little Chief (Jahns, 1954a, p. 30–34, Map 8 [Nos. 1–14], Figures 11-15; Jahns, 1954b, p. 36, Figure 5; Weber, 1963, p. 85–115, Pl. 1, 3, and references included therein). (See Figure 3 – Mines and Quarries from Pala to Mesa Grande.)

Pala Conglomerate. "Valley fill of a type not common in this area occurs in the valley of the San Luis Rey, in the vicinity of Pala. This material is a conglomeratic mass of boulders and residuum, having a thickness of about 200 feet above and extending to an undetermined depth below the present level of the river. The boulders which make up a large part of this formation are all regular, most of them being prismatic blocks with slightly rounded corners showing that they have been transported only short distances. ... This deposit is older than the valley fill which underlies the present valley floors ... but it has not been possible to ascertain definitely the relative ages of these formations" (Ellis and Lee, 1919, p. 70). Jahns (1954a, p. 35–36, Map 8, Figures 11 and 16) refers to these deposits as the Agua Tibia Fan (Figure 3), derived principally from the adjacent Agua Tibia (Palomar) Mountains.

Jahns (1954a, p. 35) notes that: "The[se] deposits contain scattered vertebrate remains of late Pleistocene age, and probably are in part correlative with the younger arkosic fill of the Elsinore-Temecula trough." It is further noted on page 36 that there is good evidence that the Agua Tibia fan formed a temporary dam across the San Luis Rey River during late Pleistocene time (note *bunched-up* contours on what would have been the former right abutment, W½, Section 31, T9S, R1W, SBBM; United States Geological Survey, 1968c). Evidence for an ephemeral lake includes abundant Quaternary age plant fossils found in fine-grained lacustrine deposits (drill-hole cuttings) near the southeastern end of Pauma Valley.

About a mile east of the community of Pala, construction aggregate has been mined on the north side of State Route 76 from an alluvial fan at the distal end of Magee Creek. This is likely the same Pala Conglomerate Formation described by Ellis and Lee (1919) and Jahns (1954a) above.

Dimension Stone. This product was mined about 3½ miles northeast of Pala (Section 19, T9S, R1W, SBBM) from 1923 until at least the late 1950s. Spheroidally weathered boulders, as well as sub-angular boulders displaced by mass wasting, were extracted from the McGee Gabbro (or "black granite") quarry (Weber, 1963, p. 260; Hoppin and Norman, 1950, p. 4, 11, 13, 15, and 18. See also Jahns, 1954a, p. 35, Map 8 [No. 15], Figures 11 and 17; and Figure 3).

Rincon Pegmatite District. Pegmatite dikes, some of which contain small quantities of beryl, kunzite, and tourmaline, lace the hillsides north and east of the community of Rincon, located at the intersection of State Route 76 and Valley Center Road. This district was prospected and mined during the first decade of the 20th century, but little more than sporadic collecting has occurred since (Hanley, 1951; Weber, 1963, p. 87, Pl. 1). See also, Jahns (1954a, p. 36–37, Map 8 [Nos. 16, 17, 18]). Most of the mines are located on the south-southwesterly flank of Rincon Mountain (Figure 3).

SEGMENT 3

Segment 3 lies between S½, Section 31, T10S, R1E, SBBM, and NE¼, NW¼, Section 33 (projected), T10S, R1E, SBBM. The upper end of this segment is at about River-Mile 41.23 and river elevation 1760 feet. Across this stretch, the river gradient steepens dramatically to about 355.55 feet per mile over a distance of about 2.25 miles.

The upper part of this segment occurs about 2 miles south-southeast of the intersection of State Route 76 and South Grade Road. Mid- to late-Cretaceous, crystalline granitic and older metamorphic rocks underlie this rather steep and narrow stretch of river canyon. It is this pronounced and steep knickpoint that separates the lower reaches of the San Luis Rey River from its upper reaches (Sauer, 1929, p. 217). River rafting, hunting, and fishing have likely comprised the main human uses over the years.

SEGMENT 4

Leaving the above-noted NE¼, NW¼, Section 33 (projected), T10S, R1E, SBBM, and continuing upstream to Henshaw Dam, the river gradient flattens to about 88.44 feet per mile over a distance of about 9.95 miles. Still traveling in a fairly narrow canyon, and underlain and bordered by granitic and metamorphic rocks of the Peninsular Ranges batholith, camping and picnic grounds can be found at relatively easy access points along the stream.

Palomar Observatory. Located about 6 miles north from this stretch of San Luis Rey River, the 200-inch Hale Telescope was installed in 1948 at the Palomar Mountain Observatory. See sidebar, next page, for a geological side trip from State Route 76 to the Palomar Observatory.

Mesa Grande Pegmatite District. Located a little over 3 miles southwest from Henshaw Dam, Mesa Grande pegmatite district mines have produced gem quality tourmaline, gem beryl, garnet and quartz (Jahns, 1954a, p. 41, Map 8 [Nos. 19–23]; Weber, 1963, p. 87, Pl. 1). The Himalaya mine is probably the most famous in this district (Weber, 1963, p. 91–92, 104, Pl. 1). See also Figure 3.

Mesa Grande Gold District. This little known gold mining district is located approximately 2½ miles east of the above noted Mesa Grande pegmatite district. The Black Eagle and Shenandoah mine workings are located in northeast trending (northwest dipping), one- to two-foot wide quartz veins imbedded in schist and mixed granitic rock. These 5-stamp-mill claims were worked in the 1880s–1890s and again briefly in the 1930s (Weber, 1963, p. 138, 163, Pl. 1). See also Figure 3.

At the toe of Henshaw Dam, the San Luis Rey River is at River-Mile 51.18 and at an elevation of 2640 feet. The 123 feet high, 650 feet wide earth-fill dam was built primarily for irrigation water storage by the Vista Irrigation District in 1922-23. Discharge is via the San Luis Rey River and the side-hill Escondido Canal to Lake Wohlford for distribution to agricultural and now mostly residential consumers.

Elsinore Fault and Henshaw Dam. The Elsinore fault (Photo 9) passes directly under Henshaw Dam (Willets, 1967, striking photo on p. 117; Evans, 1988, p. 54–59; Photo 10). Built prior to our present day understanding of faults and earthquakes, the dam was retrofitted in 1980-81 with a downstream rock-fill buttress dam and a permanently lowered spillway (Bischoff and others, 2003, p. 69–76; Photo 11).

SEGMENT 5

Leaving Henshaw Dam, the *named* upper reach tributary of San Luis Rey River backtracks northeasterly across *Valle de San Jose* (Warner Basin) en-route to *Puerta la Cruz* (at SR-79), and further on to Puerta la Cruz Conservation Camp (river elevation 3000 feet, River-Mile 60.67). The river gradient across this 9.49 mile stretch of Warner Basin is 37.93 feet per mile.

Warner Basin. This basin has been noted and described by several authors over the years, including Hanks (1886), Fairbanks (1893), Ellis and Lee (1919, p. 48, 70, Pl. 3), Sauer (1929, p. 210), California Department of Public Works (1935, p. 36–37, Map of Principal Drainage Basins), California Department of Public Works (1939), Jahns (1954a, p.

PALOMAR MOUNTAIN SIDE TRIP

0 South Grade Road beginning at State Highway 76.

1.1 Roadcut exposures of very coarse-grained alluvial-fan and landslide debris.

1.6 The highway traverses an area underlain by fault-shattered tonalite, capped in places by landslide masses.

2.2 Roadcut exposures of sheared and shattered rock in the Elsinore fault zone.

2.7 Additional exposures of severely fractured tonalite and granodiorite, covered locally by very coarse-grained slump and slide deposits.

3.2 The broad upland surface of the Mesa Grande appears to the southeast, beyond the canyon of the San Luis Rey River.

3.6 The trees of the Agua Tibia Mountains clearly show the effects of altitude on the distribution of life zones. Here mixed stands of oak and Bigcone spruce appear in the canyons at an altitude of about 4,000 feet.

4.0 Large exposures of shattered tonalite and hybrid schists and gneisses mark the major break that defines the Elsinore fault zone on the northeast.

4.7 Roadcut exposures of Bonsall tonalite and foliated rocks of hybrid origin.

5.0 Beginning of pine and cedar forest, although oak and spruce remain dominant at this level.

5.5 Roadcuts expose foliated rocks of the Julian schist (Triassic?), which dip moderately to steeply east-northeast.

6.0 Stand of small Coulter pine trees.

6.9 Summit. Turn left (north).

7.0 Junction with Ridge Road. Continue straight ahead on Observatory Road, which passes through heavy stands of incense cedar. Numerous roadcuts expose rocks of the Julian schist.

10.3 The road crosses a swale that marks the trace of the south branch of the Agua Tibia fault, a longitudinal break that lies within the range.

10.9 The road crosses the north branch of the Agua Tibia fault, whose general trace is marked in the area to the southeast by elongate shallow depressions and by aligned gulches and canyons.

11.7 Palomar Observatory. The main observatory building houses the telescope with the famous 200-inch mirror. The museum and observation gallery are open to visitors daily from 8 A.M. to 5 P.M.

Retrace route to State Highway 76.

Reference: Jahns, R. H., 1954a, Geology of Southern California: State of California, Department of Natural Resources, Division of Mines, Bulletin 170, Geological Guide No. 5, Northern Part of the Peninsular Range Province, 59 p., pages 37-38. Retrieved September 24, 2013, from http://archive.org/details/boxsouthgeology00calirich

38–43, Map 9, Figures 19 and 20), Babcock (1958), Merriam (1958, p. 14–15, Pl. 1), Begnoche (1960), Scheliga (1963), Weber (1963, p. 33, Pl. 1), Rogers (1966), Willets (1967, p. 115–118), and United States Department of Agriculture (1973).

Stripped of Cenozoic overburden, one would likely find an irregular, undulating relict bedrock surface dating back to the ancient erosion surface, or peneplain (Ellis and Lee, 1919, p. 74–76; Sauer, 1929, p. 223–231; Merriam, 1958, p. 14, 19; Scheliga, 1963, Figures 2, 4, 7a, and 7b; and Gastil, 1961, p. 1–3, Figure 1).

The most comprehensive and complete work on Warner Basin geology and groundwater resources was prepared by Scheliga (1963). Although his focus was groundwater resources, his treatment of geologic conditions and constraints was superior for a student at the master's thesis level of education.

Located in a tectonically active region of the Peninsular Ranges, the basin is surrounded by intermediate composition granitic rocks and mixed granitic and metamorphic rocks. Northwest trending strike-slip faults bound and cut through the basin. Warner Basin sediments are offset and tilted by post depositional tectonic activity (Scheliga, 1963, Figures 4, 7a, 7b; Weber, 1963, Rogers, 1966).

Two mappable, fluvial and lacustrine clastic sedimentary units were delineated and correlated with the Temecula Arkose (Scheliga, 1963, p. 24–34) and Pauba Formation (Scheliga, 1963, p. 35–40) of Mann (1951, 1955). Mann's dissertation work was done 30 miles northwest of Warner Basin, in Temecula Valley.

Temecula Arkose is compared with Warner Basin Arkose by Scheliga (1963, p. 33, Table 1) as follows:

"**TABLE 1.** *Comparison of Temecula Arkose in the type locality with the Warner Basin Arkose*

Formation Characteristics	Temecula Arkose (Temecula–Mann 1951)	Temecula Arkose (Warner Basin)
General characteristics & composition	Pale buff to white arkose beds, buff to brown silts, white to pale gray silicified algal limestone. Paucity of traceable units.	Buff to light gray arkose, light brown to green-brown silts, pale green to gray clayey silts, poorly defined bedding.
Source	Intermediate plutonic rocks. Evidence suggests tonalite.	Intermediate plutonic rocks. Probably tonalites on north and east flanks of basin.
Rhyolite tuff bed	A 4-foot bed of rhyolite tuff disturbed by faulting.	Rhyolite tuff beds exposed at 3 locations, ranging to 3 feet thick, disturbed by faulting.
Nature of basal contact	Lies unconformably upon either basement rocks or Tertiary igneous and sedimentary rocks.	Lies unconformably upon basement rocks.
Maximum thickness	Approx. 2450 feet.	Approx. 1000 feet.
Sedimentary features	Soil zones, cross bedding, iron-manganese concretions, silt boulders, calcareous crack fillings.	Cross-bedding, calcareous crack fillings, lacustrine deposits, localized calcareous cementing.
Diagenesis	Decomposition of feldspars in and below the surface soil zone.	Decomposition of feldspars in and below the surface soil zone.
Soil classification: U. S. Department of Agriculture	Ramona sandy loam	Ramona sandy loam"

Temecula Valley Pauba Formation is compared with Warner Basin Fanglomerate by Scheliga (1963, p. 41, Table 2) as follows:

"**TABLE 2.** *Comparison of Pauba Formation in the type [area] with the Warner Fanglomerate*

Formation Characteristics	Pauba Formation (Pauba-Mann 1951)	Pauba Formation (Warner Basin)
General character and composition	Coarse fanglomerate & interbedded sands & silts of highly variable composition. Much oxidized iron manifesting as red or brown soils, as limonite streaks, or as iron films around grains of quartz and feldspar. Abundant cobble & boulder beds.	Coarse fanglomerate & interbedded sands & silts of highly variable composition. Much iron oxide staining throughout formation & developed soil. Cobble and boulder beds predominant.
Source	Palomar Block & Aguanga Block.	Palomar Block, Aguanga Block & Agua Caliente Block.
Stratigraphic position	Angular & erosional unconformity with underlying Temecula Arkose.	Angular & erosional unconformity with underlying Temecula Arkose.
Nature of surface expression	Capping, relatively undissected table land. More resistant to erosion than Temecula Arkose.	Capping moderately dissected, gently sloping table land. More resistant to erosion than Temecula Arkose.
Diagenesis	Cobbles and boulders weathered in place. Igneous rocks disintegrate readily. Resistance to chemical weathering usually determines state of decomposition.	Cobbles and boulders weathered in place. Igneous rocks disintegrate readily. Metamorphic rocks are usually more resistant.
Soil classification: U. S. Department of Agriculture	Hanford sandy loam.	Hanford sandy loam."

Other than having been deposited in basins approximately 30 miles apart, the similarities suggest that these two sedimentary units likely had similar genetic origins, provenance, and tectonic afterlife.

Age of Temecula Valley and Warner Basin Sediments

Temecula Arkose

"The age of the Temecula Arkose in Warner Basin ... is probably Lower Pleistocene and correlates in time with previously mapped [Mann, 1951, 1955] Temecula Arkose." (Scheliga, 1963, p. 31, 34).

Biostratigraphic studies of Temecula Arkose in the Temecula Valley, indicate it varies from approximately 3.5 million years to 0.75 million years of age (approximately late Pliocene–mid-Pleistocene) (Pajak and others, 1996, p. 39).

Also in Temecula Valley, ash beds found within the upper unnamed sandstone member of Temecula Arkose have been correlated with the 0.785 ± 0.002 million years old Bishop ash (Kennedy, 1977b, p. 5; Pajak and others, 1996, p. 31, 39 and references therein).

Dr. Tom Rockwell (San Diego State University) confirmed the presence of Bishop Tuff ash in Warner Basin sediments (personal communication, June 13, 2013). This strongly suggests at least a temporal correlation between Ice Age deposits in Temecula Valley and Warner Basin.

"Outcrop[s] of interbedded white rhyolite tuff..." were reported by Scheliga (1963, p. 28, 29, 31) to have been found in Section 20 (projected), T10S, R3E, SBBM. In the next section to the north, an approximately 1⁄16 to ¼ inch thick lens of white ash(?) was found intercalated within a thinly bedded claystone and siltstone in a road cut on the east side of State Route 79, approximately ⅛ mile south of the turnoff to the Puerta la Cruz Conservation Camp (Section 17 [projected], T10S, R3E, SBBM; Photo 12).

Temecula Arkose in Warner Basin is exposed in numerous road cuts (Photo 13) along State Route 79 between Morettis Corner ("T" intersection with State Route 76) and Warner Springs Resort.

Pauba Formation

"No direct evidence of the age of the Pauba Formation was discovered in the subject area [Warner Basin]: however, Mann (1951) assigned the Pauba Formation to early-late Pleistocene. That age agrees favorably with the conditions of deposition and history of the formation in Warner Basin." (Scheliga, 1963, p. 40).

Biostratigraphic studies of the Pauba Formation, in Temecula Valley, indicate that the age varies from approximately 0.6 million to 0.12 million years (late Pleistocene) (Kennedy, 1977, p. 5–6; Pajak and others, 1996, p. 39).

Other than being stratigraphically younger than the underlying Temecula Arkose, biostratigraphic evidence for the age of the Pauba Formation in Warner Basin is, for the time being, apparently absent.

The Pauba Formation is exposed in road cuts along the north side of State Route 79 between Warner Springs Resort (Stage Road) and the turnoff to Indian Flats Campground (Lost Valley Road [aka Indian Flats Road, Forest Service Road 9-S-05]). One of the most dramatic Pauba Formation outcrops is located 0.43 mile north of State Route 79 on the east side of the road to Indian Flats Campground. Just before the road begins a series of tight turns, 10 to 20 feet high, near-vertical road cuts expose distinctively yellow-brown, clastic, fine-grained to gravelly sediments (Photo 14).

Although there is no direct physical correlation between Temecula Valley sediments and Warner Basin sediments, the above-noted age determinations for Temecula Valley strata may at least provide a ballpark estimate for ages of the sedimentary packages in the Warner Basin.

Author's Comment. Because of the distance that separates Temecula Valley and Warner Basin, and the likelihood that the provenance of each basin is spatially different, the Warner Basin deposits might better be given their own individual names – accompanied by a clear statement that they are likely correlative (in both space and time) with the Temecula Valley sediments. For example, the stratigraphically lower arkose unit in Warner Basin could be named the "Warner Basin Arkose," and the overlying equivalent of the Pauba Formation could be named the "Henshaw Fanglomerate." (North American Commission on Stratigraphic Nomenclature, 2005).

Small Fault North of Morettis Junction. Approximately ½ mile northerly from the Morettis Corner, a small fault (Photo 15) occurs in a road cut in Temecula Arkose on the left (west) side of the road. Just for drill, determine a quick and dirty strike, dip, and apparent offset in this outcrop. Besides the obvious evidence for apparent vertical separation, is there any good evidence for lateral offset? Before getting too involved with this exercise, turn around and look for fault evidence in the road cut on the east side of the pavement (Photo 16). This new data point will help establish a strike line, and will help verify the dip. Does this help with determining a dip-slip, vs. strike-slip, vs. oblique-slip sense of fault separation?

Sedimentary Enigma. An enigmatic geologic outcrop occurs on the east side of State Route 79, about 1.8 miles south of Morettis Junction (intersection of State Route 76 and State Route 79), and approximately 0.1 mile north of the turn off to the Santa Ysabel Casino. This approximately 1,000 feet long, 45 feet high, road cut is composed of thinly to massively bedded fine-grained sediments (with an occasional boulder thrown in just for good measure). A spring and seepage near the middle of the outcrop has caused sapping and slumping – hence the white K-Rails to keep debris off the paved highway right-of-way (Photo 17).

These sediments are about 330 feet higher in elevation than the general topographic surface elevation of Warner Basin sediments in the vicinity of Morettis Junction. Additionally, these sediments do not match the coarse-grained arkosic character of most of the Warner basin sediments.

The question then becomes, what is the origin of this enigmatic outcrop? There does not appear to be any compelling evidence to correlate it with Eocene Ballena Gravels that crop out in the vicinity of Ramona (Ellis and Lee, 1919, Pl. 1; Weber, 1963, Pl. 1). So...

- Is it somehow related to the Warner Basin sedimentary regime?
- Is it a distant cousin of the Pauba Formation?
- Is it the remains of an isolated pond created by landsliding or debris flows?
- Is it the remains of a sag pond along the Elsinore zone of faults?
- Where did the other side of the basin and its sediments go? Is it possible that they might be found somewhere along the Elsinore fault zone?

Whatever the answer, the fine-grained nature of the sediment begs for a relatively quiet-water, low-energy depositional setting.

Go ahead and give the old gray matter a couple of scratches. What do you think? How might you go about searching for the answer?

Mystery Pedestal. This mushroom-shaped rock, or tor (Neuendorf and others, 2005, p. 676), is reminiscent of other similarly weathered and sculpted granitic remnants in San Diego County's backcountry. This mysterious tor of intermediate composition granitic rock (Photo 18) occurs on the south side of Forest Service Road 9-S-05 (Indian Flats Road), about 0.95 mile north of the turn off into Indian Flats Campground.

Could the upper smooth surface represent a period of landscape stability,like the one responsible for the ancient peneplain (Sauer, 1929; Gastil, 1961)? Was the surface of this pedestal at or near the ancient ground surface at that time? Could it be that during that *quiet time,* moist soil below the ancient ground surface resulted in increased chemical weathering which would later lead to a more easily physically eroded neck? Then with a change in the balance of forces (such as climate and tectonic regime), could landscape instability have resulted in about 5 to 6 feet of general landscape denudation exposing the approximately 2- to 4-foot diameter pedestal as we see it today? Is today's landscape stable or unstable? How many mushroom-shaped cousins do you think there might be in San Diego County?

Montezuma Gold Mining District. About 12 miles east of Henshaw Dam, gold was mined in the late 1800s and early 1900s in the hills north of Montezuma Valley Road, near the north end of Old Mine Road (Weber, 1963, p. 156, Pl. 1; Clark, W. B., p. 174, Pl. 1).

Butterfield Overland Stage and Mail. This service operated from September 1857 through June 1861. Stage stops were located at Warner's Ranch (Photo 19) and at nearby Oak Grove (Photo 20). The Warner's Ranch house site is located along County Road S2, about ¾ mile east of its "T" intersection with State Route 79. Via State Route 79, Oak Grove is located about 16¾ miles northwest from the "T" intersection of County Road S2 and State Route 79. (Wikipedia, 2013—1; Wikipedia, 2013—2).

Camp Wright. During the American Civil War, the Union Army established Camp Wright in October 1861 to guard lines of communication along the wagon road between Fort Yuma, Arizona, and the California coast. Originally located at a fork in the trail at Warner's Ranch, the camp was moved in November 1861 about 17½ miles northwest to a more sheltered location (less hostile micro-climate) at Oak Grove. Camp Wright was abandoned December 1866. (Wikipedia, 2013—1; Wikipedia, 2013—3; Historical Plaque at Oak Grove).

Warner Hot Springs Resort. Prior to the arrival of European emigrants and Spanish missionaries, Native American Cupeño Indians inhabited Warner Basin. By the early- to mid-19th century the American West was in the midst of a changing human-scape — from a hunter-gatherer lifestyle to an agricultural, ranching, and mining, style of living.

John Trumbull Warner, an American-born citizen passed through this broad highland valley on his way to the west coast of California. In 1844 he became a naturalized Mexican citizen and changed his name to Juan Jose Warner. Later, he was given the broad highland valley we know today as Warner Basin or Henshaw Basin through a Mexican land grant. For many years, Warner operated Rancho San Jose del Valle as a successful cattle ranch. By mid-20th century Warner's resort and spa flourished until after World War II (Wikipedia, 2013—4). At the time of this writing, the resort was closed to the public (Photo 21).

Warner Hot Springs most likely owes its existence to the presence of the northwest trending Aguanga fault, a member of the broadly disbursed Elsinore system of faults (Scheliga, 1963, Figure 2; Rogers, 1966; and Weber, 1963, Pl. 1). In the 1950s, the author remembers that water flowed first into a *hot* swimming pool and then later into a *cool* pool.

In 1980, Higgins and Martin reported the following measurements at Warner Hot Springs.

Temperature	59° Celsius (centigrade)	138° Fahrenheit
Flow	526 liters per minute	139 gallons per minute
Total dissolved solids	350 milligrams per liter	350 parts per million

Mountain Peaks. Warner Basin, is surrounded by some of the highest peaks in San Diego County. Five of these, in order of elevation in feet above mean sea level are:

Hot Springs Mountain, 6533'. The highest peak in San Diego County, this cone-shaped peak is prominently located approximately 4 miles northeast of Warner Hot Springs Resort. It is the site of an abandoned fire lookout tower (United States Geological Survey, 1960c; Photo 22).

Bucksnort Mountain (Combs Peak), 6193'. Located approximately 8 miles north of Warner Hot Springs Resort, this unassuming high point is located near the headwaters of the San Luis Rey River (United States Geological Survey, 1960b).

High Point Lookout (Palomar), 6140'. Located on the Palomar block, about 1 mile east of Palomar Observatory, this ridgeline "pimple" is manned as a fire lookout by volunteer observers during the summer fire season (United States Geological Survey, 1949–82).

Vulcan Mountain (Pechacho Peak), 5719'. This relatively nondescript, rounded high-point is located approximately 7 miles southeast of Warner Basin (United States Geological Survey, 1960d).

Angel Mountain, 4554'. Located about 2 miles southwest of Warner Basin, this unassuming rounded high-point is located between the Mesa Grande pegmatite and Mesa Grande gold districts (United States Geological Survey, 1948; Weber, 1963). See Figure 3.

SEGMENT 6

From Puerta la Cruz Conservation Camp to the headwaters of San Luis Rey River through Johnson Canyon (at River-Mile 68½, elevation 5790 feet), the average gradient of this "bumpy" stretch is 356.83 feet per mile along the final 7.83 miles. Granitic and metamorphic rocks underlie and border this last portion of the river.

Indian Flats Campground. This campground is located on the east side of San Luis Rey River, along the common section line between sections 27 and 34, T9S, R3E, SBBM. Access is via Forest Service Road 9-S-05 (aka Indian Flats Road and Lost Valley Road), an asphalt-concrete-paved road that leaves State Route 79 at Mile Post 37.05. This delightful primitive campground was used by the South Coast Geological Society and the San Diego Association of Geologists for their joint fall 2003 field trip (Murbach and Hart, eds., 2003).

Blue Bell Tourmaline Mine. This mine is located at the northeast end of Chihuahua Valley, on the west side of Johnson Canyon (SW¼, Section 12, and NW¼, Section 13, T9S, R3E, SBBM). Discovered and active in the early 1900s, gem quality blue tourmaline is now principally of interest to mineral collectors (Weber, 1963, p. 89, 98, Pl. 1).

Road Trip along the San Luis Rey River Corridor. To view the San Luis Rey River profile in the field, start at the Pacific Ocean, River-Mile Zero (Photo 1),* and proceed along State Route 76 to its eastern end at Morettis Junction ("T" junction with State Route 79). This route crosses the river twice. The first crossing is on new twin concrete pillar bridges at about River-Mile 12. Just downstream from these new bridges, Photo 2 shows the old abandoned early 20th century concrete arch bridge. The second crossing (identified only by corrugated metal guard rails painted bright white on both sides of the road) is at River-Mile 50.2, about a mile downstream from Henshaw Dam.

For the remainder of this adventure, use the Warner Springs (1959), Beauty Mountain (1960a), and Bucksnort Mountain (1960b) 7.5' topographic quadrangle maps as basic guides through this seeming maze of backcountry roads.

At the end of State Route 76, turn left (north) on State Route 79 and proceed approximately 7.8 miles past Warner Springs Resort, past the turn off to Indian Flats Campground, past the glider-port, and past the turnoff to Puerta la Cruz Conservation Camp — to a modern concrete bridge at Mile Post 39.95. Lush greenery on the up- and down-stream sides of the bridge stand out in stark contrast to the surrounding, gently rolling hills of normally dry brown grass. This is the *named* San Luis Rey River, upstream from Lake Henshaw.

---

* See photo data at end of paper for GPS coordinates.

Turn around and go back approximately 2.8 miles, past the glider-port, and turn left (north) at Mile Post 37.05 on Forest Service Road 9-S-05 (aka Indian Flats Road and Lost Valley Road). Follow this paved road 6.0 miles to the right-hand (west) turnoff to Indian Flats Campground. Follow the pavement into camp and go to the far west end of the camp loop road and park. From here it is a short hike west-southwesterly to the San Luis Rey River and the cataracts and falls — a delightful sight and worth a photo or two! Don't stop short and miss the large rounded and smoothed plunge pools carved in the granitic bedrock by centuries of swirling sand and debris-laden winter runoff water falling and cascading down the steep slopes. This is akin to Nature's original sandpaper.

Leave the campground and turn left (north) on the dirt truck trail. Continue 0.95-mile to the Mystery Pedestal (Photo 18) located on the south side of the road, and another 0.1-mile to the concrete low-water bridge crossing of San Luis Rey River (Photo 23). A high-clearance, four-wheel drive vehicle is recommended for this poorly-maintained but passable stretch of road. Beyond this river crossing the unimproved truck trail north to Chihuahua Valley is in VERY BAD SHAPE! It is *not* recommended! Enjoy the adventure and fresh mountain air, take some photographs and return to State Route 79. That is about it for viewing the river by car — anything else will require a horse, bicycle, or boots on the ground. There is a lot of private property in this area, so a word to the wise, inquire before you expire.

DISCUSSION AND CONCLUSIONS

At this reconnaissance level of investigation, the principal take-home message from this brief look at the San Luis Rey River longitudinal profile is that the knickpoint, just east of Harrah's Rincon Casino, attests to uplift of the Peninsular Ranges and disruption of the previously graded profile. The elevation change across about 2¼ miles is about 800 feet.

As to timing, it is surely post Peninsular Ranges batholith intrusion and subsequent uplift and erosion that is well-documented by Late Cretaceous sediments shed to, and deposited in, basins along the present-day Pacific Ocean coastline — or post about 66 million years ago. It may also be post Eocene (post about 34 million years ago), the approximate age of the Ballena Gravels that document rivers flowing east to west, across peneplain. These auriferous fluvial deposits have been largely eroded away to a few residual hilltop remnants that are still visible south of Ramona, California. (Ellis and Lee, 1919, Pl. 3; Weber, 1963, Pl. 1; Rogers, 1966; Kennedy and Peterson, 1975; Kennedy, 1977a, Kimbrough and others, 2001.)

Without additional data points to work with, I think my favorite Multiple Working Hypotheses (Chamberlin, 1890) guess is that this uplift event may coincide with the approximately 6 million years ago opening of the Gulf of California-Salton Trough, and relocation of the Pacific North American plate transform boundary from its ocean home to its present location along the east side of the Salton Trough (Atwater, 1970). What do you think? Guesses, guesstimates, prognostications, and suggestions are welcome. How many hypotheses, no matter how outrageous, can you add to the list?

POSTSCRIPT

For you insomniacs, here is some homework to keep your left brain busy at 3am.

An investigation into weather and climate, specific rock types and faulting was beyond the scope of this cursory look at the San Luis Rey River longitudinal profile. So, for those of you who are interested in doing further library research and field work, these topics could keep you busy after work, on weekends, and in between the infamous honey dos!

Additionally, the following projects should keep you out of the bars and jails for at least a field-season or two.

1. Radiometric age dating rhyolite tuff in the Warner Basin Arkose.
2. Establish a biostratigraphic correlation of Warner Basin sediments with nearby strata, for example, in Temecula Valley.
3. Determine origin and history of the enigmatic sedimentary outcrop on the east side of State Route 79, just north of the Santa Ysabel Casino turnoff.
4. Investigate a possible correlation between mystery pedestal rocks and landscape cycles of stability and instability.
5. Delineate and characterize the deeply eroded, hummocky mass of quartz diorite that crops out between Warner Basin and Chihuahua Valley. This approximately five miles long by one mile wide, northwest-southeast trending irregularly-shaped geomorphic surface is roughly traversed by Forest Service Road 9-S-05 (Indian Flats Road). San Luis Rey River runs diagonally through it. Is it possible that a correlation exists between this ancient surface and the nearby abandoned fan surface underlain by Pauba Formation? And, is there any relationship with the Temecula Arkose?
6. Now to the origin of Warner Basin itself. Is it part of a *porpoise structure* along the northeast side of the Elsinore fault, sandwiched between Palomar Mountain and Vulcan Mountain? Are there other explanations? Is there a similar *porpoise cousin* on the southwest side of the Elsinore fault?
7. Do knickpoints, similar to that described herein for San Luis Rey River, exist in other nearby major coastal drainages, for example, Santa Margarita, San Dieguito, San Diego, Sweetwater, Otay and Tijuana rivers? If so, could these be related and correlated to a common origin?
8. And, with a little imagination and ingenuity, there are a number of senior theses, and even master's theses projects, in the San Luis Rey River watershed that could be addressed to make significant contributions to science.

ACKNOWLEDGEMENTS

I would like to thank Brian J. Olson for encouraging me to pursue this reconnaissance investigation and Dr. Monte Marshall for helpful and stimulating discussions. Many, many thanks go to Dr. Pat Abbott, Dr. Kerry Cato, Dr. Vic Camp, Mr. Richard Proctor, Dr. Bob Reynolds, Dr. Tom Rockwell, and Dr. Stephen Schellenberg, all of whom assisted me with timely responds to pleas for help finding older and otherwise obscure references. Both Dave Bloom and Carole Ziegler assisted with final smoothing of the manuscript and illustrations. As usual, my patient wife, Wendy, has endured yet another *endless* project.

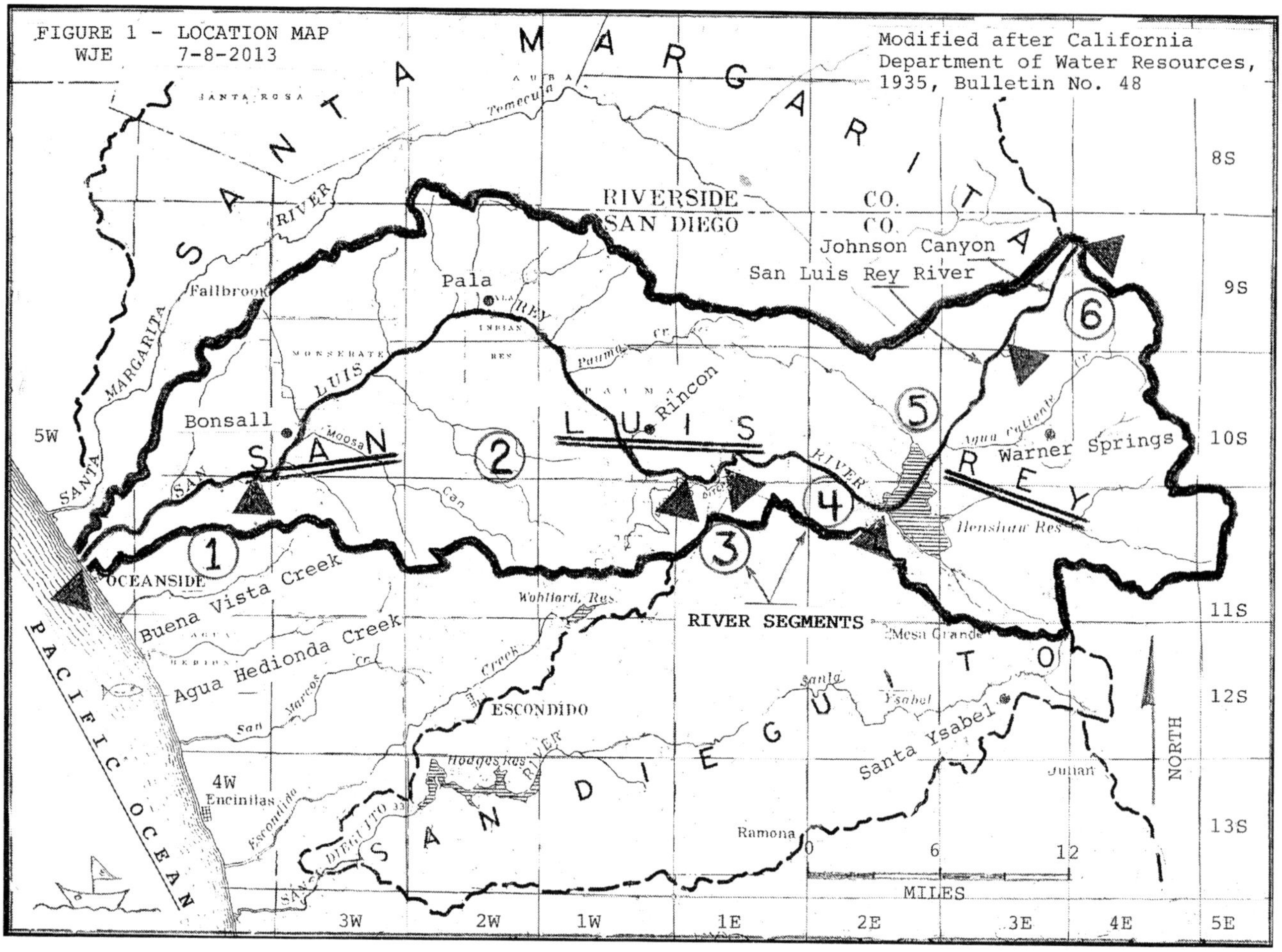

Figure 1. *Location Map.*

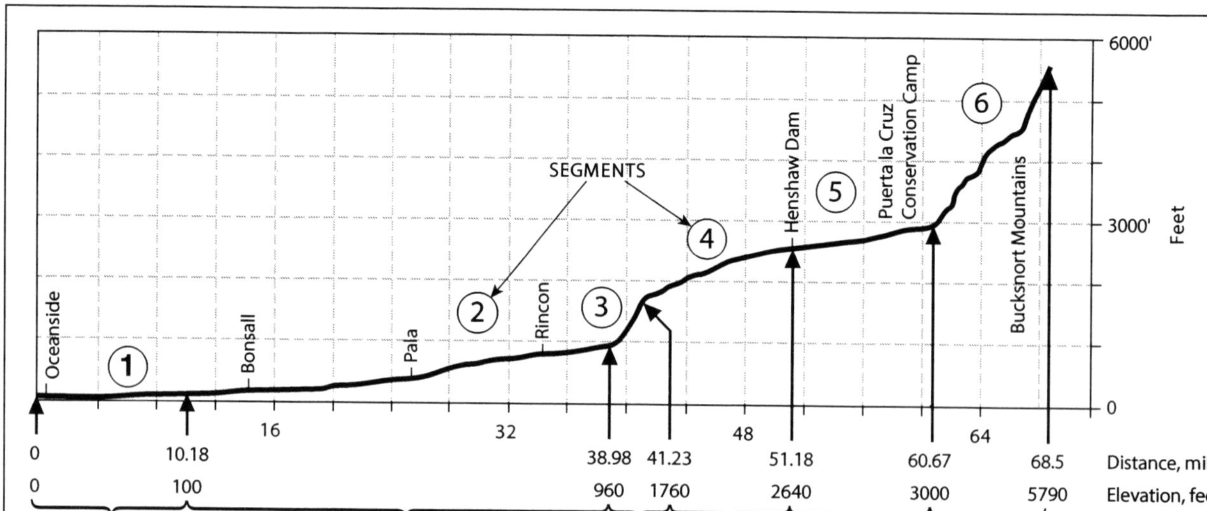

	1	2	3	4	5	6	Segment
	10.18	28.8	2.25	9.95	9.49	7.83	Length, miles
	100	860	800	880	360	2790	Fall, feet
	9.8	29.86	355.55	88.44	37.93	356.83	Gradient, feet/mile
	Tertiary Sediments	Granitic and Metamorphic (Segments 2–6)					Geology

Figure 2. *San Luis Rey River Longitudinal Profile and Segment Data Summary. Profile taken from U.S.G.S. 7½′ Topographic Quadrangle Maps, References, and Figure 1 – Location Map. Horizontal to vertical exaggeration: 21.12 to 1.*

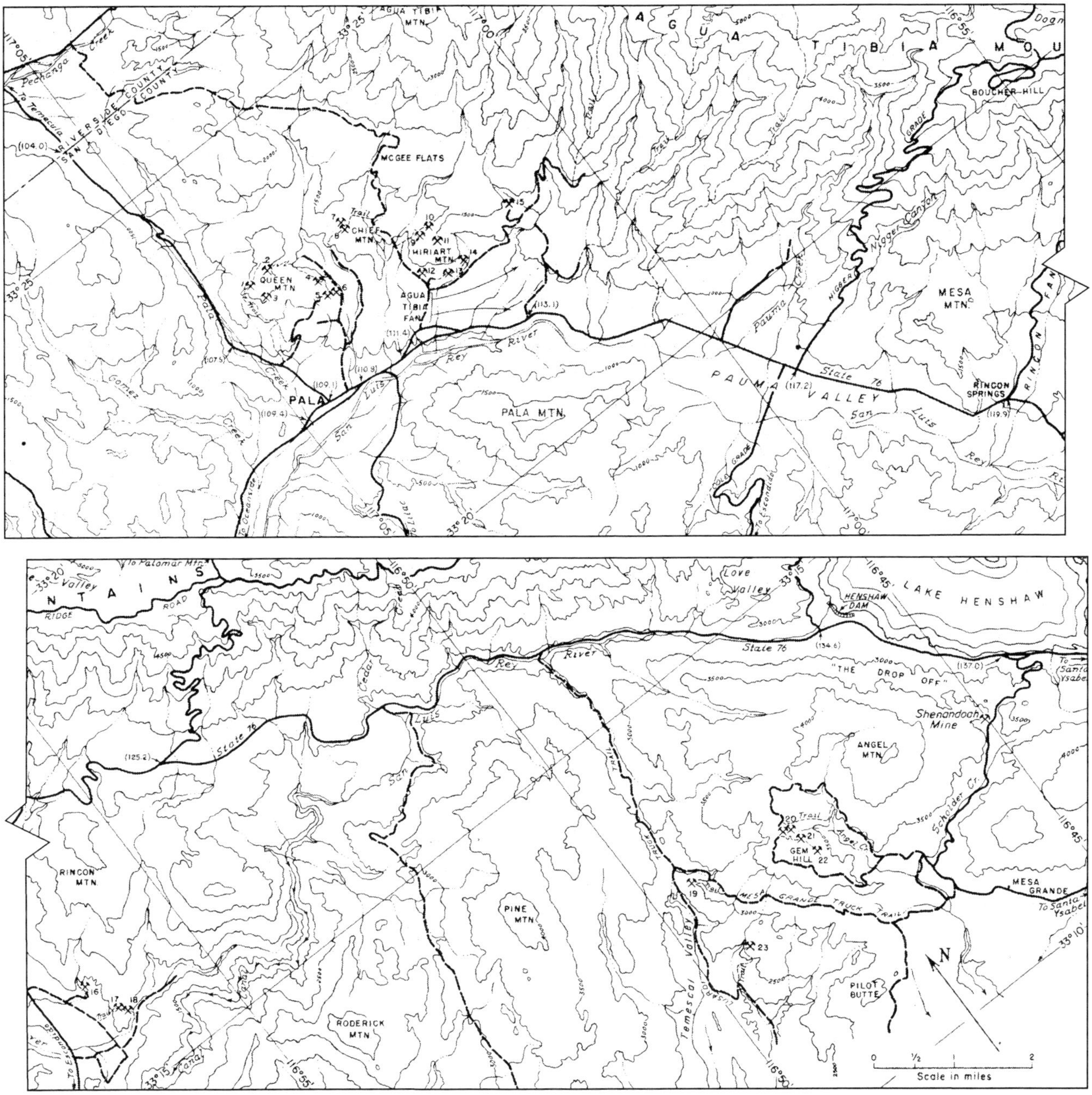

Figure 3. *Mines and Quarries from Pala to Mesa Grande (Jahns, 1954a, Map 8 [left half above, right half below.]). Map 8. Pala to Warner Basin and Mesa Grande. Numbered mines and quarries are as follows: (1) Tourmaline King—tourmaline, quartz. (2) Tourmaline Queen—tourmaline, quartz. (3) White Cloud—tourmaline, quartz. (4) Gem Star—tourmaline, quartz. (5) Stewart—lepidolite, ambilygonite, tourmaline. (6) Douglass—tourmaline, quartz. (7) Ocean View—beryl, quartz. (8) Pala Chief—spodumene, tourmaline, beryl, quartz. (9) Senpe—beryl, lepidolite, quartz, tourmaline. (10) San Pedro—beryl, spodumene, quartz, tourmaline. (11) Vanderburg—spodumene, beryl, quartz, tourmaline. (12) Katerina—spodumene, lepidolite, quartz, beryl, tourmaline. (13) Fargo—tourmaline, quartz. (14) El Molino—beryl, quartz. (15) Johnson (McGee)—dimension stone. (16) Mack—beryl. (17) Victor—beryl, tourmaline, spodumene, garnet. (18) Clark—quartz, beryl. (19) Esmeralda—tourmaline, beryl, quartz. (20) Himalaya—tourmaline, beryl, quartz. (21) San Diego—tourmaline, quartz. (22) Mesa Grande—tourmaline, beryl, quartz. (23) Rose Quartz—quartz, garnet.*

PHOTO DATA				
Photo Number	**Description**	**Longitude***	**Latitude***	**Elevation* meters & feet**
1	San Luis Rey River Mouth	117° 23′ 19.27″W	33° 12′ 10.67″ N	2m, 6.5′
2	Old River Road Bridge	117° 14′ 13.80″W	33° 15′ 35.08″ N	46m, 151′
3	Proposed Gregory Canyon Land Fill Site	117° 06′ 45.69″W	33° 21′ 32.11″ N	123m, 404′
4	Pala Mission Bell Tower circa 1900	117° 04′ 27.67″W	33° 21′ 54.88″ N	126m, 413′
5	Pala Mission Bell Tower circa 1953	117° 04′ 27.78″W	33° 21′ 55.04″ N	126m, 413′
6	Pala Mission Bell Tower July 2, 2013	117° 04′ 28.34″W	33° 21′ 55.08″ N	126m, 413′
7	Pala Mission Chapel	117° 04′ 27.60″W	33° 21′ 55.72″ N	126m, 413′
8	Stuart Lithia Mine	117° 03′ 45.18″W	33° 22′ 52.97″ N	324m, 1063′
9	Elsinore Fault at Henshaw Dam, East Grade Road	116° 45′ 50.23″W	33° 14′ 31.14″ N	839m, 2753′
10	Warner Basin, Lake Henshaw & Elsinore Fault	116° 43′ 09.45″W	33° 12′ 32.90″ N	831m, 2726′
11	Henshaw Dam	116° 45′ 54.49″W	33° 14′ 27.77″ N	829m, 2720′
12	Ash Bed(?), SR-79, Section 17	116° 41′ 30.97″W	33° 18′ 11.93″ N	904m, 2966′
13	Temecula Arkose, SR-79	116° 41′ 57.88″W	33° 13′ 46.36″ N	855m, 2805′
14	Small Fault (Elsinore Fault Zone, west side SR-79	116° 42′ 40.98″W	33° 12′ 29.02″ N	839m, 2753′
15	Small Fault (Elsinore Fault Zone), east side SR-79	116° 42′ 40.56″W	33° 12′ 28.38″ N	839m, 2753′
16	Pauba Formation, Indian Flats Road , FSR 9-S-05	116° 39′ 30.79″W	33° 17′ 32.18″ N	940m, 3084′
17	Santa Ysabel Casino Landslide SR-79	116° 41′ 28.21″W	33° 10′ 42.09″ N	941m, 3087′
18	Mushroom Rock, south side of FSR 9-S-05	116° 39′ 52.45″W	33° 21′ 27.64″ N	1162m, 3814′
19	Warner Ranch House, S2	116° 39′ 57.44″W	33° 14′ 32.02″ N	884m, 2900′
20	Butterfield Stage Station / Camp Wright, Oak Grove, SR-79	116° 47′ 41.01″W	33° 23′ 24.93″ N	831m, 2726′
21	Warner Springs Ranch, SR-79	116° 38′ 03.27″W	33° 16′ 59.58″ N	955m, 3133′
22	Hot Springs Mountain, Former Fire Lookout, SR-79	116° 39′ 08.28″W	33° 16′ 15.34″ N	937m, 3074′
23	San Luis Rey River, Low Water Crossing, FSR 9-S-05	116° 39′ 57.30″W	33° 21′ 28.84″ N	1152m, 3780′

*Except for aerial images, Longitude, Latitude, and Elevation are at the approximate camera location.

Figure 4. *Google Earth GPS Coordinates for Photos 1–23.*

Photo 1. *San Luis Rey River Mouth (Photo by W. J. Elliott, 004-a, 7-4-2013).*

Photo 2. *Old State Route 76 Bridge at North end of East Vista Way, Vista, California (Photo by W. J. Elliott, 018-a, 7-4-2013).*

Photo 3. *Proposed Gregory Canyon Landfill Site (Photo by W. J. Elliott, 052(89)-a, 7-2-2013).*

FIGURE 9. The Mission Asistencia at Pala, as it appeared shortly after the turn of the present century. *Photo courtesy of Mrs. Frank A. Salmons.*

Photo 4. *Detached Bell Tower at Pala Mission, circa 1900 (Jahns, 1954a, p. 28; 005).*

FIGURE 10. The detached bell tower of the reconstructed Pala mission, as it appears today.

Photo 5. *Detached Bell Tower at Pala Mission, circa 1953 (Jahns, 1954a, p. 29; 006).*

Photo 6. *Detached Bell Tower at Pala Mission, July 2013 (Photo by W. J. Elliott, 052(84)-a, 7-2-2013).*

Photo 7. *Pala Mission Chapel, July 2013 (Photo by W. J. Elliott, 052(86)-a, 7-2-2013).*

Photo 8. *Northwest Aerial Oblique View of Stewart Tourmaline/Lepidolite Mine (Jahns, 1954a, p. 33; 009).*

Photo 9. *Elsinore fault in Road Cut Northwest of Henshaw Dam (Photo by W. J. Elliott, 052(82)-a 7-2-2013).*

Elsinore Fault Zone Traversing
Unit 3, January 1966

The San Luis Rey River is interrupted by Lake Henshaw, one of the largest water storage areas in the Region.

Photo 10. *Northwest Aerial Oblique View of Warner Basin, Lake Henshaw, and Elsinore Fault (Willets, D. B., 1967, p. 117; 004).*

Photo 11. *Retrofitted Lake Henshaw Dam (Photo by W. J. Elliott, 052(81)-a, 7-2-2013).*

Photo 12. *Thin White Ash(?) Seam in State Route 79 Road Cut, Section 17 [projected], T10S, R3E, SBBM (Photo by W. J. Elliott, 052(71)-a, 7-2-2013).*

Photo 13. *Temecula Arkose Road Cut Approximately 2¼ Miles North of "T" Junction of State Route 79 and State Route 76 – Morettis Junction (Photo by W. J. Elliott, 052(14)-a, 7-2-2013).*

Photo 14. *Pauba Formation Outcrop, East Side of Road to Indian Flats Campground, Approximately 1/3-Mile North of State Route 79 (Photo by W. J. Elliott, 052(51)-a, 7-2-2013).*

Photo 15. *Small Fault in Temecula Arkose on West Side of State Route 79. Road cut is approximately ½-mile north of "T" Junction of State Route 79 and State Route 76 – Morettis Junction (Photo by W. J. Elliott, 052(10)-a, 7-2-2013).*

Photo 16 (opposite). *Small Fault in Temecula Arkose on East Side of State Route 79. Road Cut is approximately ½-mile north of "T" Junction of State Route 79 and State Route 76 – Morettis Junction (Photo by W. J. Elliott, 052(13)-a, 7-2-2013).*

Photo 17. *Landslide, Spring, and Seepage on East Side of State Route 79, 0.1 Mile North of Entrance to Santa Ysabel Casino (Photo by W. J. Elliott, 052(07)-a, 7-2-2013).*

Photo 18. *Mystery Pedestal, North of Indian Flats Campground (Photo by W. J. Elliott, 052(42)-a, 7-2-2013).*

Photo 19. *Warner Ranch House (Photo by W. J. Elliott, 052(18)-a, 7-2-2013).*

Photo 20. *Oak Grove Barn (Butterfield Stage Stop and Camp Wright) (Photo by W. J. Elliott, 052(32)-a, 7-2-2013).*

Photo 21. *Warner Springs Ranch Resort (Photo by W. J. Elliott, 052(76)-a, 7-2-2013).*

Photo 22. *Hot Springs Mountain (Peak, 6,533') (Photo by W. J. Elliott, 052(24)-a, 7-2-2013).*

Photo 23. *Concrete Low-Water Culvert-Crossing over San Luis Rey River, One Mile North of Turn Off to Indian Flats Campground (Forest Service Road 9-S-05) (Photo by W. J. Elliott, 052(41)-a, 7-2-2013).*

SELECTED REFERENCES

Atwater, T., 1970, Implications of Plate Tectonics for the Cenozoic Tectonic Evolution of Western North America: Geological Society of America Bulletin, v. 81, p. 3513–3536.

Babcock, B., 1958, Ground Water Occurrence and Quality, San Diego County: Unpublished Master's Thesis, University of Southern California, 123p.

Begnoche, D. J., 1960, Geology and Ground Water of Henshaw Basin: Unpublished term paper, San Diego State College, 36p.

Bischoff, J. A., MacDonald, T. C., and Wilson, T. M., 2003, Rehabilitation of an Old Hydraulic Fill Dam for Stability Against Earthquakes, in Murbach M. L., and Hart, M. W., eds., Geology of the Elsinore Fault Zone, San Diego Region: Joint San Diego Association of Geologists and South Coast Geological Society annual field trip guidebook, Sunbelt Publications, San Diego, California, 209p.

California Department of Natural Resources and Soil Conservation Service, United States Department of Agriculture, February, 1952, Know California's Land.

California Department of Public Works, 1935, Publications of the Division of Water Resources, E. Hyatt, State Governor: San Diego County Investigation, Bulletin No. 48.

California Department of Public Works, 1939, Division of Water Resources, Net Annual Water Yield from Henshaw Reservoir, 67p.

Chamberlin, T. C., 1890, The Method of Multiple Working Hypotheses: Science, v. 15, p. 92-96.

Clark, M. K., Maheo, G., Saleeby, J., and Farley, K. A., 2005, The Non-Equilibrium Landscape of the Southern Sierra Nevada, California: Geological Society of America Today, v. 15, n. 9, p. 4–10.

Clark, W. B., 1970, Gold Districts of California: California Division of Mines and Geology, Bulletin 193, 186p.

Elliott, W. J., and Marshall, M., 2012, Lawrence Canyon Fault, Oceanside, California, in Bloom, D. and Gomes, C., eds., Waiting for Tsunami: Coastal Hazards of Northern San Diego County, California: San Diego Association of Geologists annual field trip guidebook, September 8 and 9, 2013, Sunbelt Publications, San Diego, California, p. 119–134.

Ellis, A. J., and Lee, C. H., 1919, Geology and Ground Waters of the Western Part of San Diego County, California: United States Geological Survey Water-Supply Paper 446, 321p., approximate map scale (Pl. 3) 1″ = 4 miles.

Evans, J. R., 1988, Henshaw Dam, in Evans, J. R., ed., Landslides in Crystalline Basement Terrain, San Diego Association of Geologists annual field trip guidebook, 87p.

Fairbanks, H. W., 1893, Geology of San Diego — also portions of Orange and San Bernardino Counties: California Mining Bureau Report 11, p. 76-120.

Gallen, S.F., Wegmann, K. W., and Bohnenstiehl, D. R., 2013, Miocene Rejuvenation of Topographic Relief in the Southern Appalachians: Geological Society of America Today, v. 23, n. 2, p. 4–10.

Gannett, Henry, 1901, Profiles of Rivers in the United States: United States Geological Survey, Water Supply Paper No. 44, 100p., 11 Pl.

Gastil, R. G., 1961, The Elevated Erosion Surfaces, in Thomas, B. E., ed., Field Trip Guidebook prepared for the 57th Annual Meeting of the Cordilleran Section of the Geological Society of America at San Diego State College [University], p. 1–3, Figure 1.

Hanks, H. G., 1886, San Diego County: California Mining Bureau Report 6, Part 1, p. 80–90.

Hanley, J. B., 1951, Economic Geology of the Rincon Pegmatites, San Diego County, California: California Division of Mines, Special Report 7-B, 24p., Pl. 1, map scale 1:24,000, contour interval 40′.

Higgins, C. T., and Martin, R. C., 1980, Geothermal Resources of California: California Division of Mines and Geology, Geologic Data Map No. 4, map scale 1:750,000.

Hinds, N. E. A., 1952, Evolution of the California Landscape: California Division of Mines, Bulletin 158, 240p. (see also, Jenkins, 1938.)

Hoppin, R. A., and Norman, L. A., Jr., 1950, Commercial "Black Granite" of San Diego County, California: California Division of Mines, Special Report 3, 19 p.

Hubbard Scientific, Inc., 1979, Santa Ana Sheet, 1° x 2° plastic raised relief topographic map number NI-11-8, scale 1:250,000, contour interval 200′ (Hubbard Scientific, Inc., Chippewa Falls, WI.).

Hyatt, Edward (State Engineer), 1935, San Diego County Investigation: California Division of Water Resources, Bulletin 48.

Jahns, R. H., 1954a, Geology of Southern California: State of California, Department of Natural Resources, Division of Mines, Bulletin 170, Geological Guide No. 5, Northern Part of the Peninsular Range Province, 59 p.

Jahns, R. H., 1954b, Geology of the Peninsular Range Province, Southern California and Baja California, in Jahns, R. H., ed., California Division of Mines, Bulletin, 170, Chapter 2, Contribution 3, p. 29–52.

Jenkins, O. P., 1938, Geomorphic Map of California: California Division of Mines, Bulletin 158, Pl. 2, map scale 1:2,000,842, contour interval 2,000′.

Kennedy, M. P., and Peterson, G. L., 1975, Geology of the San Diego Metropolitan Area, California: California Division of Mines and Geology Bulletin 200, 56 p., map scale 1:24:000, contour interval 20'.

Kennedy, M. P., 1977a, Geology of National City, Imperial Beach and Otay Mesa Quadrangles, Southern San Diego Metropolitan Area, California: California Division of Mines and Geology Map Sheet 29, map scale 1:24,000, contour interval 20'.

Kennedy, M. P., 1977b, Recency and Character of Faulting along the Elsinore Fault Zone in Southern Riverside County, California: California Division of Mines and Geology, Special Report 131, 12 p., Pl. 1, map scale 1:24,000, contour interval 20'.

Kimbrough, D. L., Smith, D. P., Mahoney, J. B., Grove, M., Gastil, R. G., Ortega-Rivera, A., and Fanning, C. M., 2001, Forearc-Basin Sedimentary Response to Rapid Late Cretaceous Batholith Emplacement in the Peninsular Ranges of Southern and Baja California: Geology, June 2001, v. 29, n. 6., p. 491–49.

Larsen, E. P., Jr., 1948, Batholith and Associated Rocks of Corona, Elsinore, and San Luis Rey Quadrangles, Southern California: Geological Society of America Memoir 29, June 21, 1948, 182 p., map scale 1"=2 miles.

Larsen, E. P., Jr., Everhart, D. L., and Merriam, R., 1951, Crystalline Rocks of Southwestern California: California Division of Mines, Bulletin 159, 136 p., map scale 1:62,500, contour interval 50'.

Leet, D. L., and Judson, S., 1958, Physical Geology: Prentice-Hall, Inc., Englewood Cliffs, New Jersey (second edition, sixth printing, 1962), 502p. (Longitudinal profile – p. 162–163 and figure 10-5 shows 3 longitudinal profiles and reference to Gannett, H., USGS Water-Supply Paper 44, 1901.)

Longwell, C. R., Knopf, A., and Flint, R. F., 1932, A Textbook of Geology – Part I, Physical Geology: John Wiley & Sons, Inc., London, 514p. (p. 45 Longitudinal profile, p. 46 Figure 20 – longitudinal profile of Lehigh River.)

Mann, J. F., Jr., 1951, Late Cenozoic Geology of a Portion of the Elsinore Fault Zone, Southern California [PhD dissertation]: University of Southern California, 163p.

Mann, J. F., Jr., 1955, Geology of a Portion of the Elsinore Fault Zone, California: California Division of Mines, Special Report 43, 22p., Pl. 1, map scale 1"= 1 mile, contour interval 250'.

Merriam, R. H., 1958, Geology and Mineral Resources of Santa Ysabel Quadrangle, San Diego County, California: California Division of Mines, Bulletin 177, 42p., Pl. 1., map scale 1;62,500, contour interval 100'.

Merriam, R. H., and Bischoff, J. I., 1975, Bishop Ash – A Widespread Volcanic Ash Extended to Southern California: Journal of Sedimentary Petrology, v. 45, n. 1, p. 207–211.

Miller, W. J., 1935, Geomorphology of the Southern Peninsular Range of California: Geological Society of America Bulletin, v. 46, p. 1535–1562.

Moyle, W. R., Jr., 1971, Water Wells in the San Luis Rey River Valley Area, San Diego County, California: California Department of Water Resources, Bulletin No. 91-18, 347p.

Murbach, M. L., and Hart, M. W., eds., 2003, Geology of the Elsinore Fault Zone, San Diego Region: Joint San Diego Association of Geologists and South Coast Geological Society annual field trip guidebook, Sunbelt Publications, San Diego, California, 209p.

Neuendorf, K. K. E., Mehl, J. P., Jr., and Jackson, J. A., eds., 2005, Glossary of Geology, Fifth Edition: American Geological Institute, Alexandria, Virginia, 779p.

Norris, R. M., and Webb, R. W., 1976, Geology of California: John Wiley and Sons, Inc., New York, 365p. (Peninsular Ranges – Geomorphology p. 178-79 notes post-Nevadan orogeny [mid- late-K] erosion surface – peneplain – and subsequent uplift and stepped erosional treads – for example knickpoints like seen along the San Luis Rey River at Rincon.)

North American Commission on Stratigraphic Nomenclature, 2005, North American Stratigraphic Code (final acceptance, July 2005): American Association of Petroleum Geologists Bulletin, v. 89, n. 11 (November 2005), DOI:10.1306/07050504129, p. 1547–1591. (Accessed on the web 6-23-2013.)

Pajak, A. F., III., Scott, E., and Belt, C. J., 1996, A Review of the Biostratigraphy of Pliocene and Pleistocene Sediments in the Elsinore fault Zone, Riverside County, California: PaleoBios, v. 17, nos. 2–4, p. 28–49.

Perazzo, Peggy and George, last updated 7-4-12, Quarries and Beyond: http://quarriesandbeyond.org/states/ca/quarry_photo/ca-san_diego_photos_1_b.html (Accessed on the web 8-7-2013).

Rogers, T. H., compiler, 1966, Geologic Map of California – Santa Ana Sheet: California Division of Mines and Geology, map sheet, scale 1:25,000, contour interval 200'.

Sarna-Wojcicki, A. M., Bowman, H. R., Meyer, C. E., Russell, P. C., Woodward, M. J., McCoy, G., Rowe, J. J., Jr., Baedecker, P. A., Asaro, F., Michael, H., 1984, Chemical Analysis, Correlations, and Ages of Upper Pliocene and Pleistocene Ash Layers of East-Central and Southern California: United States Geological Survey Professional Paper 1293, p. 1–39.

Sauer, C., 1929, Land Forms in the Peninsular Range of California as Developed about Warner's Hot Springs and Mesa Grande: University of California Publications in Geography, v. 3, n. 4, p. 199–290.

Scheliga, J. T. Jr., 1963, Geology and Water Resources of Warner Basin, San Diego County, California [M.A. thesis]: University of Southern California, 91p., map scale 1:24,000, contour interval 100'.

Strand, R. G., compiler, 1962, Geologic Map of California – San Diego-El Centro Sheet: California Division of Mines and Geology, Map Sheet, map scale 1:250,000, contour interval 200'.

Thomas, B. E., ed., 1961, Field Trip Guidebook, San Diego County – 1961: Prepared by the Geology Department at San Diego State College [University] March, 1961, for the 57th Annual Meeting, Cordilleran Section, Geological Society of America and Associated Societies, 71p.

Thornbury, W. D., 1969, Principals of Geomorphology: John Wiley & Sons, Inc., New York (second edition), 594p. (p. 106 The Graded Stream, p. 108 Longitudinal profile, p. 135–136 An Idealized Fluvial Cycle, Youth, Maturity, Old-Age.)

United States Department of Agriculture, 1973, Soil Survey – San Diego Area, California: U.S.D.A., Parts I and II, map scale 1:24:000.

United States Geological Survey, 1948, Mesa Grande, California, 7½' topographic quadrangle map, scale 1:24,000, contour interval 20.

United States Geological Survey, 1948-82, Boucher Hill, California, 7½' topographic quadrangle map, scale 1:24,000, contour interval 40'.

United States Geological Survey, 1949-82, Palomar Observatory, California, 7½' topographic quadrangle map, scale 1:24,000, contour interval 40'.

United States Geological Survey, 1959, Warner Springs, California, 7½' topographic quadrangle map, scale 1:24,000, contour interval 40'.

United States Geological Survey, 1959-69, Santa Ana topographic map, 1° x 2° sheet, scale 1:250,000, contour interval 200'.

United States Geological Survey, 1960a, Beauty Mountain, California, 7½' topographic quadrangle map, scale 1:24,000, contour interval 40'.

United States Geological Survey, 1960b, Bucksnort Mountain, California, 7½' topographic quadrangle map, scale 1:24,000, contour interval 40'.

United States Geological Survey, 1960c, Hot Springs Mountain, California, 7½' topographic quadrangle map, scale 1:24,000, contour interval 40'.

United States Geological Survey, 1960d, Ranchita, California, 7½' topographic quadrangle map, scale 1:24,000, contour interval 40'.

United States Geological Survey, 1960e, Warners Ranch, California, 7½' topographic quadrangle map, scale 1:24,000, contour interval 40'.

United States Geological Survey, 1968a, Bonsall, California, 7½' topographic quadrangle map, scale 1:24,000, contour interval 20'.

United States Geological Survey, 1968b, Morro Hill, California, 7½' topographic quadrangle map, scale 1:24,000, contour interval 20'.

United States Geological Survey, 1968c, Pala, California, 7½' topographic quadrangle map, scale 1:24,000, contour interval 20'.

United States Geological Survey, 1968-75a, Oceanside, California, 7½' topographic quadrangle map, scale 1:24,000, contour interval 20'.

United States Geological Survey, 1968-75b, San Luis Rey, California, 7½' topographic quadrangle map, scale 1:24,000, contour interval 20'.

Weber, F. H., Jr., 1963, Geology and Mineral Resources of San Diego County, California: California Division of Mines and Geology, County Report 3, 309p., map scale 1" = 2 miles.

Weber, F. H., Jr., 1982, Recent Slope Failures, Ancient Landslides and Related Geology of the North-Central Coastal Area, San Diego County, California: California Division of Mines and Geology, Open-File Report 82-12-LA, 77p., Pl. 1, map scale 1:24,000, contour interval 20'.

Willets, D. B., 1967, Ground Water Occurrence and Quality – San Diego Region [California], Volume I – Text: California Department of Water Resources, Bulletin No. 106-2, 235p., 10 plates.

Wikipedia, 2013 — 1, Oak Grove Butterfield Stage Station [includes information on Camp Wright].

Wikipedia, 2013 — 2, Butterfield Overland Mail.

Wikipedia, 2013 — 3, American Civil War.

Wikipedia, 2013 — 4, Warner's Ranch.

Giant Bison *(Bison latifrons)* Facts

TAXONOMY AND NOMENCLATURE

Describer (Date): C. H. Smith 1827 for *Bison*
R. Harlan 1825 for *Bison latifrons,* Fauna Americana: (Philadelphia: A. Finley). p. 273

Kingdom: Animalia
Phylum: Chordata
Class: Mammalia
Order: Artiodactyla (Even-toed hoofed animals: includes pigs, sheep goats, cattle, deer)
Family: Bovidae (Cattle, water buffalo, bison, antelopes, goats, sheep and more)
Genus: Bison
Species: *Bison priscus* (extinct Steppe Bison)
Species: *Bison latifrons* (extinct Giant Bison)
Species: *Bison antiquus* (extinct Ancient Bison)
Species: *Bison bison* American Bison
Subspecies: *Bison bison bison* (American Plains Bison)
Subspecies: *Bison bison athabascae* (American Wood Bison)
Species: *Bison bonasus* (European Bison)
Subspecies: *Bison bonasus bonasus* (Lowland Bison)
Subspecies: *Bison bonasus caucasicus* (extinct in 1925)
Subspecies: *Bison bonasus hungarorum* (extinct Hungarian Bison)

Taxonomic History and Nomenclature

- "Latifrons" comes from Latin words referring to a wide forehead
- First fossil bison described in North America was a *Bison latifrons* found in Kentucky (Peale 1803)
- Bison taxonomy underwent a long period of taxonomic splitting that resulted in some 10 fossil species recognized in North America by Skinner and Kaisen in 1947.
    - This taxonomy in great need of revision (McDonald 1981)
- Relationship of modern American bison and European bison is unclear at present, but both are quite similar genetically and can interbreed (Prusak et al 2004)
- Modern North American bison have two recognized subspecies: the American Plains Bison *(B. b. bison)* and the American Wood Bison *(B. b. athabascae*). (McDonald 1981)

Phylogeny

- Even-toed hoofed mammals trace their ancestry back to at least 45 million years ago in Eocene times.
- The *Bison* genus first appeared in southern Asia, around 2 million years ago (McDonald 1981)
- *Bison priscus* was the ancestor of at least some of the North American bison (Prusak 2004)
    - This species flourished in northern Eurasia and Alaska and may have been the dominant hoofed mammal there (Guthrie 1970)
- Bison migrated to North America several times in the Pleistocene during the low sea levels when exposed land connected North America and Asia
- *B. latifrons* appeared less than 200,000 years ago in North America.
- *B. latifrons* became extinct around 20,000 years ago (McDonald 1981)
- All bison nearly became extinct in North America at the end of the Pleistocene when much of the other megafauna did become extinct (McDonald 1981)

Reference: San Diego Zoological Society factsheet on long-horned and ancient bison. http://library.sandiegozoo.org/factsheets/_extinct/bison_extinct/bison_extinct.htm

Discovery of remains of an extinct giant bison (*Bison latifrons*) in Pleistocene (Rancholabrean) fluvial strata in the San Luis Rey River Valley, San Diego County, California, USA

Deméré, T.A. Department of Paleontology
Randall, K.A. San Diego Natural History Museum
Riney, B.O. P.O. Box 12390, San Diego, CA
Siren, S.A.

ABSTRACT

Pleistocene fluvial and lacustrine deposits are preserved in patches along the San Luis Rey River Valley between Interstate 5 and Interstate 15 in northern San Diego County, California, USA. Over the past 20 years, a series of residential, commercial, and public roadway construction projects along the river valley have resulted in the discovery of "Ice Age" vertebrate fossils in these Pleistocene deposits. In April 2013, freeway improvements by Caltrans to the State Route 76—Interstate 15 interchange near Pala Mesa unearthed a partial skeleton of a bison. Recovered fossils include a nearly complete skull and an articulated series of thoracic, lumbar, and sacral vertebrae, as well as both pelvic bones. An isolated right humerus and radius-ulna found 38 meters south of the other bones may be part of the same individual as are a cluster of cervical vertebrae found 52 meters north of the skull. Preliminary analysis of morphometric skull data suggests a taxonomic assignment to the extinct giant bison species, Bison latifrons. *The fossils were found near the base of a 7.6 meter thick sequence of Quaternary fluvial strata in a light yellowish gray, silty, very fine-grained, micaceous sandstone with calcrete nodules. An irregular erosional and bioturbated contact separated the fossiliferous sandstone from an overlying dark yellowish orange to brown, very fine-grained sandstone. The age of the Pleistocene strata has not yet been determined, but the occurrence of* Bison *indicates correlation with the Rancholabrean North American Land Mammal Age and a maximum age of about 220 ka. This discovery represents the first record of extinct giant bison in San Diego County and serves to expand our understanding of the paleobiogeographic range of this charismatic member of the extinct Pleistocene megafauna. Previous records of* Bison *from San Diego County have been based on fragmentary remains that are either not identifiable to species level or have been assigned to the more diminutive species,* Bison antiquus.

INTRODUCTION

Previous paleontological and geological work on the Pleistocene history of western San Diego County has focused on the marine record as preserved on the stair step sequence of emergent marine terraces (abrasion platforms) and adjacent marine and estuarine embayments that characterize the geomorphology of the coastal plain (Kern and Rockwell, 1992). These landforms and their associated sedimentary deposits are in turn the result of both global fluctuations in sea level over the past 2.5 million years, as well more regional episodes of local tectonic uplift. However, an important part of the regional Pleistocene history has been largely overlooked, namely that of the non-marine (fluvial and lacustrine) deposits that accumulated in the coastal river valleys upstream from the Pleistocene bays and estuaries. These

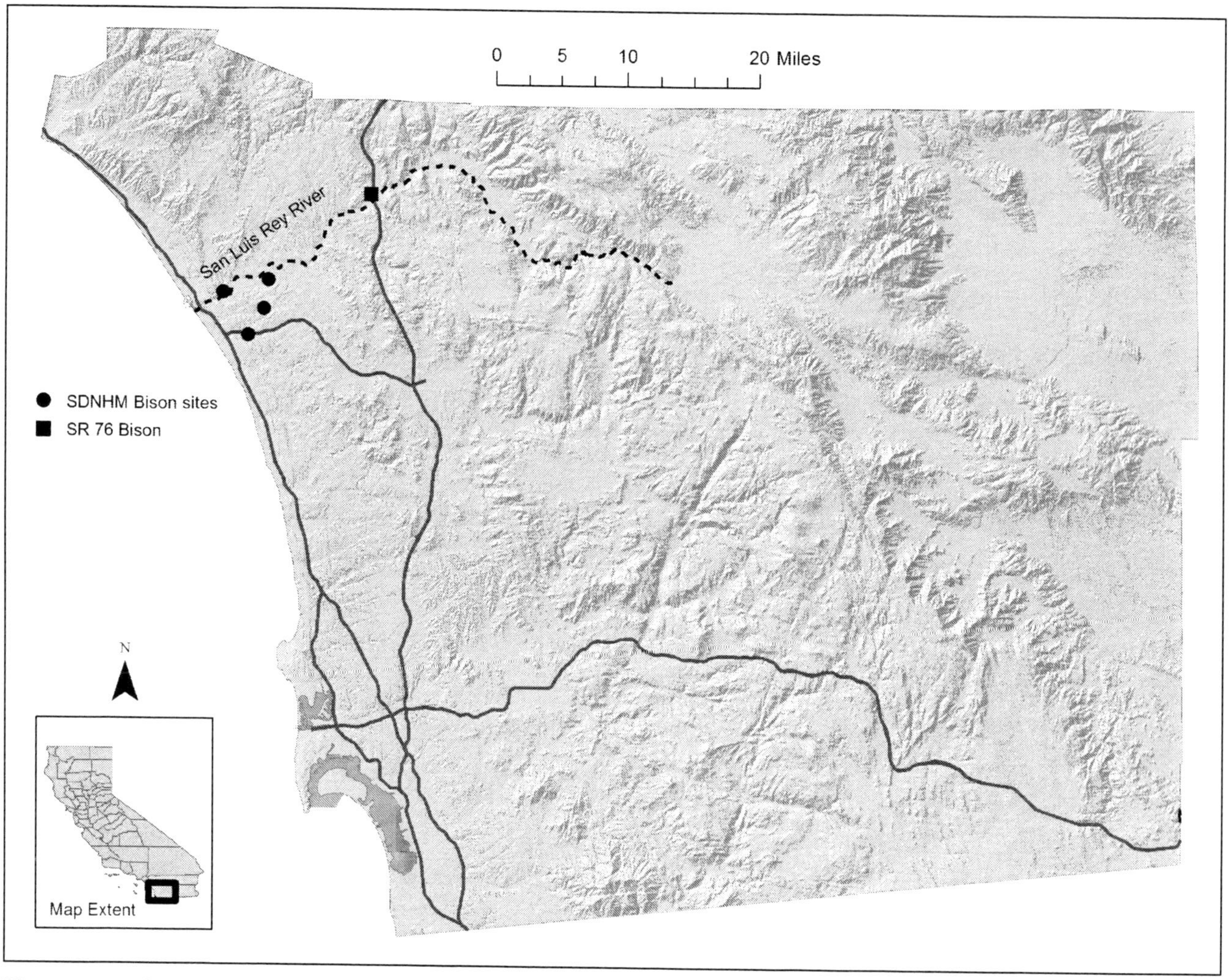

Figure 1. *Index map of San Diego County showing known records of* Bison *fossils in the San Luis Rey River Valley and adjacent coastal drainages.*

deposits are quite extensive, especially in the larger river valleys such as the Sweetwater River Valley in the south county, San Dieguito River Valley in the mid county, and San Luis Rey River Valley in the north county. Majors (1993) described a relatively diverse vertebrate assemblage (Glen Abbey local fauna) from upper Pleistocene (Wisconsin?) deposits preserved in the Sweetwater River Valley and Guthrie (2012) described a large avifauna from upper Pleistocene lacustrine strata in the San Luis Rey River Valley. The fossils described from both sites were discovered as a result of construction-related excavations and were recovered as part of paleontological resource mitigation projects conducted by paleontologists from the San Diego Natural History Museum (SDNHM). Similar mitigation activities carried out in the fall of 2013 during construction improvements to the State Route 76 and Interstate 15 interchange in the eastern part of the San Luis Rey River Valley near Pala Mesa (Figure 1) resulted in the discovery of a nearly complete skull with preserved horn cores and postcanine dentition of a giant bison (*Bison latifrons*). An intact series of sacral, lumbar, and posterior thoracic vertebrae found in articulation with the left and right pelvic bones was associated with the skull. Additional bison postcrania discovered at the site, including portions of the right forelimb and a series of cervical vertebrae, are likely from the same individual. The Pala Mesa discoveries are the most complete *Bison* remains yet found in San Diego County and represent the first record of giant bison in this region. The description and preliminary analysis of these fossils is the subject of this report.

Previous Work

Geology. The State Route 76—Interstate 15 interchange project area lies within the northern portion of the Peninsular Ranges Geomorphic Province, which is primarily composed of Lower Cretaceous (~125 to 90 Ma) plutonic rocks of the Peninsular Ranges Batholith overlain by a westward thickening wedge of Upper Cretaceous and Cenozoic sedimentary deposits (Walawender, 2000).

According to published geologic maps (Tan, 2000), the geologic units underlying the project area include, from oldest to youngest: undivided tonalite (Cretaceous age), granodiorite of Indian Mountain (Cretaceous age; ~120 Ma), tonalite of Couser Canyon (Cretaceous age; ~120 Ma), older alluvial flood plain deposits (Pleistocene age; ~500 to 10 ka), active alluvial flood plain deposits of late Holocene age (<5 ka), and active wash/stream deposits of late Holocene age (<5 ka).

Paleontology. The bovid lineage that includes modern species of the genus *Bison* originally evolved in Europe during the late Pliocene and following an initial radiation in Eurasia, eventually dispersed during the middle to late Pleistocene into North America via the Bering land bridge (McDonald, 1991). Although the taxonomy of the original *Bison* immigrant to North America is somewhat controversial, there is a consensus that once in the New World this lineage radiated to produce at least two endemic taxa, eventually leading to the modern plains bison, *Bison bison*. The extinct North American endemic taxa include the ancient bison, *Bison antiquus* and the giant bison, *Bison latifrons*.

Fossils of *Bison latifrons* and *Bison antiquus* have been reported from numerous late Pleistocene sites in North America. Both species are bi-coastal with fossil occurrences reported from California, Florida and the Eastern Seaboard. *Bison antiquus* sites have dense concentrations in California, Florida, and the central plains and north-central United States. Numerous sites have also been discovered in Mexico and in Canada, with the densest higher latitude concentration in Alberta. *Bison latifrons* fossil sites are mostly restricted to the United States, with concentrations in California, Idaho, Florida, and the Central Plains. Fossils of this species in California cluster in southern California, the San Francisco Bay Area, and the Modoc Plateau (McDonald, 1981).

Fossil remains of *Bison* have been reported from 82 discovery sites in southern California including two localities from Santa Barbara County, one from Ventura County, 45 from Los Angeles County, six from San Bernardino County, 16 from Orange County, five from Riverside County, four from San Diego County, and three from Imperial County (Jefferson, 1991; McDonald, 1981; Miller, 1968; Springer et al., 2009, 2010; Scott and Cox, 2008). Several of these localities have produced multiple bison specimens and in some cases more than one named species. The degree of taxonomic certainty regarding the specific identity of the *Bison* fossil remains also varies, with 16% of reported specimens assigned to *B. antiquus*, 10% assigned to *B. latifrons*, 10% provisionally assigned to *B. antiquus*, 5% provisionally assigned to *B. latifrons*, 57% assigned only to the genus *Bison*, and 2% provisionally assigned to *Bison*. In addition, the documented localities span a range of geologic ages that includes much of the Late Pleistocene (200 ka to 10 ka).

In San Diego County, previous reports of *Bison* are restricted to the northwestern portion of the county, within 20 miles of the coast. Most reported occurrences are from the San Luis Rey River drainage, with isolated occurrences along smaller drainages to the south including Loma Alta Creek and Buena Vista Creek (Figure 1). *Bison* fossils recovered from these northern San Diego County localities were all discovered as a result of paleontological monitoring of construction-related excavation activities. A partial right dentary fragment with a second lower molar (SDSNH 77800) was collected in August 1999 from a Rancho Del Oro residential project site in central Oceanside on the south side of SR-76 and assigned to *Bison antiquus*. In south-central Oceanside, near the intersection of Oceanside Boulevard and College Boulevard, associated skeletal elements (i.e., vertebrae, pelvic bones, and limb bones) of *Bison antiquus* (SDSNH 83253-83257) were discovered in November 2000 during mass grading of the Ocean Ranch Corporate Centre. In June 2002, a right dentary of a juvenile *Bison antiquus* with deciduous second through fourth premolars (SDSNH 101449) was collected during construction of the Summit at Carlsbad apartment complex in northern Carlsbad, along Marron Road and south of SR-78. In September 2002, construction of the Wanis View Estates residential subdivision on the north side of the San Luis Rey River Valley near the Oceanside Municipal Airport, unearthed a left dentary with deciduous third and fourth premolars, and an erupting first molar, of a juvenile *Bison* sp. (SDSNH 90056). Also collected on this project was the proximal half of a left metatarsal of an adult *Bison* sp. (SDSNH 89548). Until now, no record of giant bison (*B. latifrons*) has been reported from San Diego County.

MATERIALS & METHODS

Field Methods

General. The bison fossil was discovered as a direct result of paleontological field monitoring of earthmoving activities associated with construction of the new freeway interchange. This monitoring involved close examination of active mass grading operations (e.g., grading to construct on ramps, off ramps, and roadways), as well as inspection of deep trenching operations (e.g., trenching to install underground wet and dry utilities). Field observations were focused on the recognition of unearthed fossil remains, but also involved collection of relevant geologic, stratigraphic, and taphonomic contextual data.

Fossil Salvage. Mass grading operations associated with construction of the northbound on ramp to I-15 from eastbound SR-76 in the southeastern portion of the new freeway cloverleaf was actively monitored during the spring of 2013 and on the afternoon of April 10, 2013, SDNHM paleontologist Brad Riney discovered several smears of fossil bone clustered across a three by ten foot area at the surface of the active scraper cut. Grading operations were diverted to other areas of the construction project while the extent of the discovery was evaluated. The first fossil remains to be identified were pelvic elements and associated sacral, lumbar, and thoracic vertebrae. On the following day, additional SDNHM paleontologists (Nicole Anderson, Gino Calvano, and Sarah Siren) joined Brad Riney to begin the recovery process (Figure 2). While the pelvic girdle and associated vertebrae were prepared for plaster jacketing and removal (see below), investigation continued on the other bone smears discovered eight feet northwest of the initial discovery. This investigation revealed first a molar tooth and then a nearly complete skull. Field efforts then shifted back to recovery of the pelvic and vertebral remains. The process began by exploring around the partially exposed fossils with hand tools (e.g., small picks and shovels) to determine the lateral and vertical extent of the remains. Care was taken to leave a three to six

Figure 2. *San Diego Natural History Museum paleontologists (Nicole Anderson, Gino Calvano, and Brad Riney) recovering fossil of* Bison latifrons *from the northbound onramp to Interstate 15. Note continuing mass grading operations in background.*

inch "halo" of intact sedimentary matrix around the fossils to keep them from crumbling. At the same time a consolidant (e.g., Butvar-acetone solution) was applied to the exposed remains to stabilize broken and/or fragile bones. Eventually, the surrounding sedimentary rock was removed until all that remained was an isolated block of sedimentary rock with the enclosed fossil bones. The block was still attached at its base to the underlying stratum by a pedestal of undisturbed sedimentary rock. The pedestal was then slightly undercut to form an overhanging lip. Damp newsprint was then placed on the upper surface of the block. A solution of 20-minute Plaster-of-Paris was then mixed and 5- to 10-inch-wide strips of burlap cloth were soaked in the plaster solution. The strips were laid across the matrix block to dry in various orientations and depending upon the size of the block, one, two, or three layers of plaster-soaked burlap "bandages" were formed on each block. The blocks were then reinforced with wooden splints. Once the plaster hardened, the supporting pedestal was undercut and the block was turned over. Hand digging tools were then used to remove any excess matrix from the bottom of the block. A plaster and burlap cap was then constructed on the inverted bottom of the block, using the same method as described above. When all layers of plaster were dry and hardened, the completed plaster "jacket" was then labeled with a field number and north arrow, and removed from the construction zone.

It should be mentioned that while recovery of the skull, pelvis, and vertebrae was going on, earthwork operations were proceeding in other areas of the construction project. This earthwork required monitoring by SDNHM paleontologists and during the process additional fossils were discovered including an isolated humerus and radius-ulna and a series of five vertebrae. These bones were found in the same stratum that produced the other fossil remains, and were later determined to also represent remains of *Bison*.

Laboratory Methods

Analysis. Skull and postcranial morphologic terminology follows Skinner and Kaisen (1947) and McDonald (1981) and primarily focuses on the morphology of the horn cores and features of the dorsal surface of the skull. Linear and curvilinear measurements were taken of skull and postcranial skeletal elements using metric anthropometers with 0.1 cm increments for larger elements (i.e., >30 cm) and metric calipers with 0.1 mm increments for smaller elements. A series of standard morphometric measurements was taken following the procedures outlined in Skinner and Kaisen (1947) and McDonald (1981) and include the following: Skull Measurement (SM)-1 Spread of horn cores, tip to tip; SM-3 Horn core length, upper curve, tip to burr; SM-5 Horn core straight line distance (chord), dorsal side, tip to burr; SM-14 Least width of frontals, between horn cores and orbits; SM-15 Greatest width of frontals, at orbits; SM-OP Distance, nuchal line to tip of premaxilla; SM-ON Distance, nuchal line to nasal-frontal suture; and SM-21 Angle of divergence of horn cores, forward from sagittal plane. Skull measurements are summarized in Table 1.

In addition to skull measurements, three linear measurements each were taken from the humerus and radius following the standard methodology of McDonald (1981). These include his humeral measurement (hm)-1, approximate rotational (functional) length; hm-2, anteroposterior diameter of diaphysis at transverse minimum; hm-3, transverse minimum at diaphysis; radius measurement (rm)-1, approximate rotational (functional) length; rm-2, anteroposterior diameter of diaphysis at transverse minimum; and rm-3, transverse minimum at diaphysis. These measurements are summarized in the text.

Abbreviations used in this report include the following: SDNHM-San Diego Natural History Museum; SDSNH-San Diego Society of Natural History (the scientific organization that operates SDNHM); ka-kilo annum (1,000 years old); Ma-mega annum (1,000,000 years old).

Table 1. *Skull measurements for SDSNH 135465. Numbers in parentheses correspond to standard measurements in McDonald (1981).*

SKULL: STANDARD MEASUREMENTS	mm
minimum horn core spread (1)	1150
minimum horn core length (3)	490
minimum straight line (5)	460
anteroposterior diameter of horn core base (12)	151
least width of frontals (14)	365
greatest width of frontals (15)	422
Distance, nuchal line to tip of premaxillae (OP)	616
Distance, nuchal line to nasal-frontal suture (ON)	243
rostral width, left doubled (Skinner & Kaisen #16)	114
Width of skull and masserteric (Skinner & Kaisen #16)	182
width of nasal and maxilla contact	91
minimum width of nasal	101
anterior width of nasals	103
posterior width of nasals	120
maximum length of nasals, left	256
length of maxilla-nasal suture, left	50
length of maxilla-nasal suture, right	47
anteroposterior diameter of orbit (right side)	111
angle of divergence of horn core, left	74°
angle of divergence of horn core, right	75°
orbit angle of divergence from midline, left	51°
orbit angle of divergence from midline, right	55°

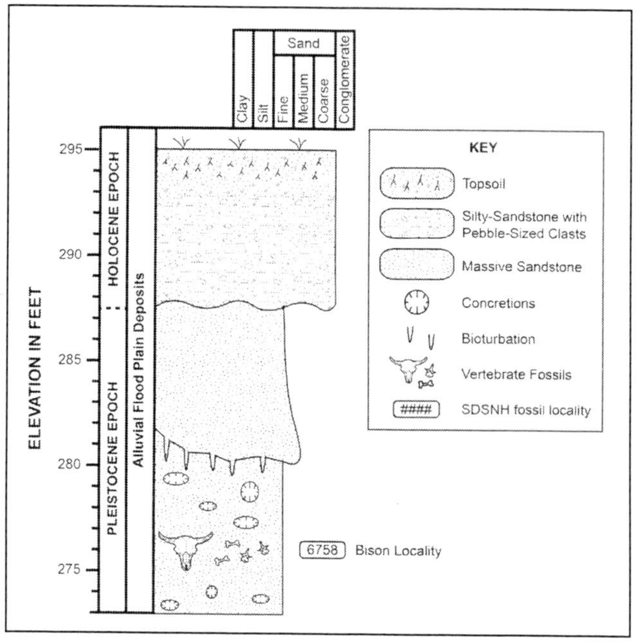

Figure 3. *Quaternary stratigraphic section exposed in the vicinity of the State Route 76-Interstate 15 interchange in northern San Diego County.*

RESULTS

Stratigraphy. The fossils were found near the base of a 25 foot thick sequence of Quaternary sedimentary rocks in a light yellowish gray, silty, very fine-grained, micaceous sandstone with calcrete nodules (Figure 3). An irregular erosional contact separated the fossiliferous sandstone from an overlying dark yellowish orange to brown, very fine-grained sandstone. The erosion surface was faintly bioturbated, with burrows extending into the fossiliferous sandstone. The fine-grained Pleistocene fluvial strata were capped by a yellowish brown, coarse-grained, pebbly sandstone of presumably Holocene age. The depth to crystalline basement at the discovery location is unknown. However, a series of three geotechnical test borings drilled along the SR-76 alignment within 90 meters of the centerline of I-15 penetrated up to 40 meters of interbedded siltstones and silty sandstones with minor beds of claystone and sandstone (Caltrans, District 11 field data). In contrast, geotechnical test borings drilled approximately 455 meters east and west of the centerline of I-15 penetrated only 2 to 3 meters of colluvial sands and residual soils before reaching igneous bedrock.

Taphonomy. The skull was lying upside-down with the snout oriented almost directly west and the horn cores roughly aligned north-south. The posterior axial skeleton and articulated pelvic bones were lying approximately 3 meters east-southeast of the skull and were dorsal side up, with the longitudinal axis of the vertebral column oriented north-south and the anterior-most vertebra positioned to the north. The right humerus and right radius-ulna were found approximately 38 meters southeast of the skull as separate elements about 6 meters apart. The cervical vertebrae and anterior thoracic vertebra were found as a disarticulated group approximately 52 meters north of the skull. Thus, the bison fossils reported here were distributed over an area of approximately 1000 m^2 and over a linear distance of approximately 71 meters. All of these

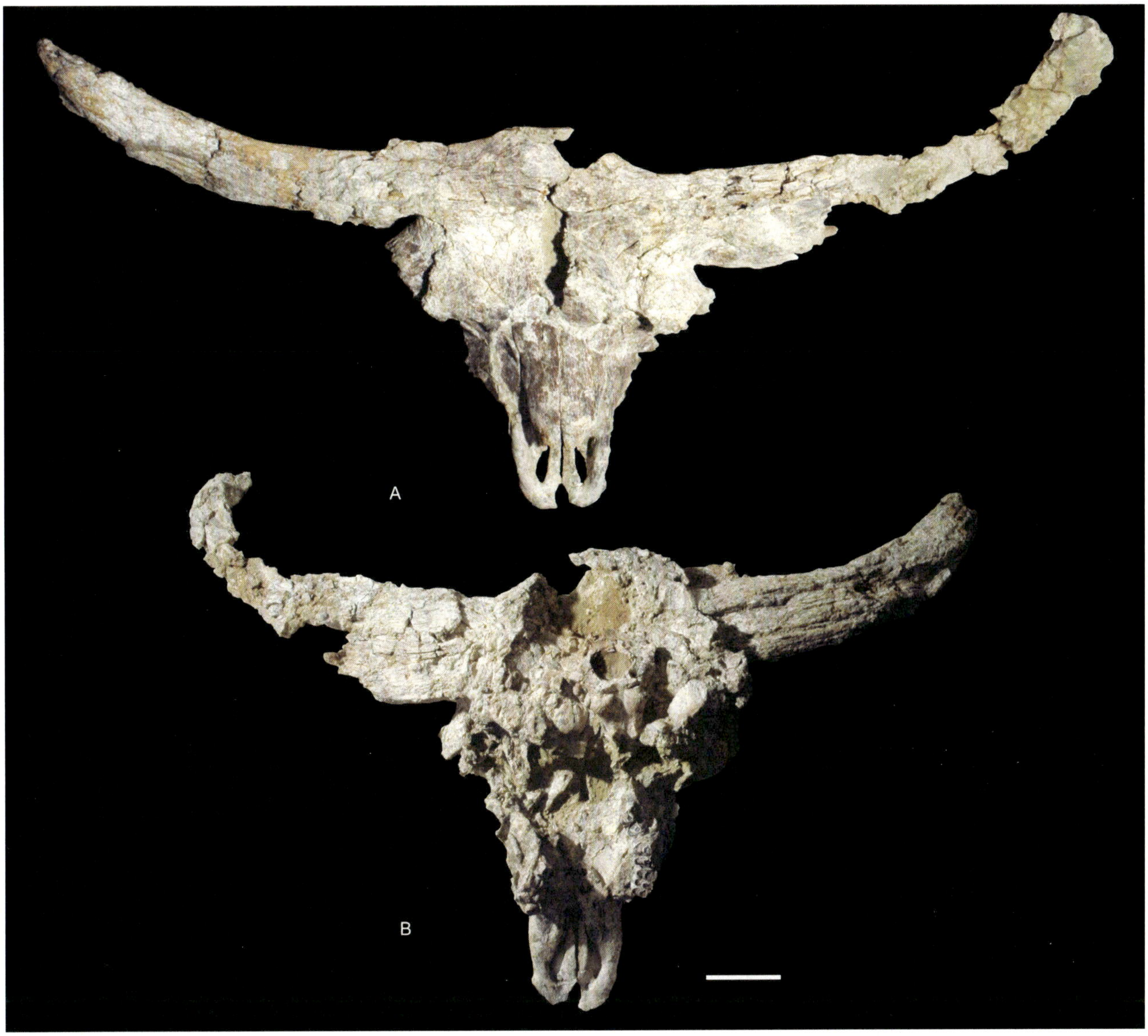

Figure 4. Bison latifrons *skull (SDSNH 135465) in dorsal (A) and ventral (B) views. Scale bar equals 10 cm.*

specimens were found in the same stratum and may be from the same individual. The remainder of the skeleton including the mandible, mid-thoracic vertebrae, ribs, scapulae, left forelimb, and right and left hind limbs were not found.

TAXONOMY & DESCRIPTION

Class Mammalia
Order Artiodactyla
Family Bovidae
Genus *Bison*
Bison latifrons (Harlan, 1825)
Figures 4, 5, 6

Referred specimens: SDSNH 135465; skull and associated postcrania (three posterior thoracic, five lumbar, and five sacral vertebrae plus right and left innominates). SDSNH 135466; associated right humerus and radius-ulna. SDSNH 135958, associated cervical (C-2, C-3?, C-4?, and C-5?) and thoracic (T-1?) vertebrae. All presumably from the same mature adult animal. Collected by SDNHM field crew, April 10-12, 2013 from SDSNH Locality 6685; SR-76/I-15 interchange, San Diego County, California, USA.

DESCRIPTION

Skull, General. The skull of SDSNH 135465 is nearly complete (Figure 4) and preserves an intact right horn core and partial left horn core (the ventral portion of the horn core was scraped away by heavy equipment). The dorsal surface of the skull is essentially complete with all major bones preserved including the premaxilla, maxilla, nasals, lacrimals, and frontals. The skull roof is deformed, with portions of the left frontal separated from the right frontal at the midline. Portions of the left frontal have been broken and offset as small, but distinct fragments. The orbital portion of the left frontal is relatively undeformed, while the right orbital portion of the frontal has been pushed medially beneath the adjacent dorsal surface of the frontal. The horn cores appear to have been posteriorly rotated, such that the horn tips and axis of curvature are directed posteriorly rather than posterodorsally. The ventral surface of the skull has been broken transversely such that the basicranium is crushed up against the region of the internal nares and rotated to the right. Related to this deformation, the ventral margin of the occipital shield has been pushed forward causing some backward rotation of at least the right horn core. The articular portions of the occipital condyles have been scraped away, leaving only a roughly circular opening (foramen magnum) into the braincase.

Premaxilla. The left and right premaxillae are undeformed and tightly articulated with the adjacent maxillae. The ascending process of the premaxilla extends as a slender bone along the anterior 60% of the lateral margin of the narial fossa, but does not contact the anterolateral corner of the nasals. This is also the condition in *Bison bison*, while in *Bos taurus*, the ascending process of the premaxilla forms the entire lateral margin of the narial fossa and extends posterodorsally to contact the nasals. The premaxillae on the palate is marked by large, right and left incisive foramina, each with a roughly spatulate outline.

Maxilla. The left maxilla is relatively undeformed, with its lateral surface primarily oriented in the vertical plane. This contrasts with the condition in most bovids where the lateral surface of the maxilla has a distinct ventrolateral slope from the nasals to the alveolar margin. The right maxilla is broken into a series of small plates of bone that ventrally become more and more displaced from each other. Like the left side, the overall slope of the lateral surface of the right maxilla is vertical and not ventrolateral. A prominent infraorbital foramen measuring 13.9 mm in diameter is positioned at the level of P3 on the left maxilla. Posteroventral to this foramen the area of the maxillary tuberosity has been broken along a longitudinal fracture line. The right maxillary tuberosity is better preserved, but also has some distortion. Although broken with varying degrees of offset, the maxilla on the palate retains a useful amount of morphology. Between the tooth rows the maxilla is as wide as it is long. The right toothrow consists of P4 through M3, while the left toothrow consists of only P2-3 (Figure 5). The molars are short anteroposteriorly relative to their transverse width and when compared with teeth of *B. bison* are less selenodont.

Lacrimal. The left lacrimal is still in articulation with the adjacent maxilla and nasal anteriorly, while posteriorly it is separated from the frontal and broken deflected laterally in the transverse plane. The right lacrimal is better preserved and remains in articulation with all adjacent bones. It measures ~141 mm in length and has the overall shape of an elongate parallelogram The nasal-lacrimal suture measures ~60 mm long. It appears not to invade between the nasal and the maxilla nearly as far anteriorly as in *Bos taurus*.

Frontal. The right frontal is largely undeformed, while the medial portion of the left frontal has been broken into several large plates of bone along roughly transverse hinge fracture lines. Individual plates are positioned at various angles from horizontal to steeply inclined. The overall deformation of the left frontal has been one of anteroposterior compression, such that the orbital portion of the frontal has been telescoped forward to override the orbital regions of the lacrimal and jugal. The midline suture between the right and left frontals is open and appears to be unfused, a condition

Figure 5. *Partial right upper dentition (P4-M3) of* Bison latifrons *(SDSNH 135465). Anterior is to the right. Scale bar equals 2 cm.*

thought to be related to ontogenetic immaturity (McDonald, 1981). The skull roof formed by the right frontal preserves a relatively flattened transverse profile, a feature reported as characteristic of specimens of *Bison latifrons* (McDonald, 1981). This contrasts with the condition in short-horned bison like *B. bison bison*, wherein the medial portion of the frontal is slightly to distinctly domed (McDonald, 1981). The orbital regions of both frontals are relatively intact and preserve features related to orbital protrusion and orbital orientation. Orbital protrusion (OP) as defined by McDonald (1981) is quantified using two metrics, absolute OP which is the maximum distance the posterior orbit margin extends lateral to the posterolateral margin of the frontal and relative OP which is expressed as a percentage of minimal frontal width (SM-14) to maximum frontal width (SM-15). In SDSNH 135465 these values are 28 mm and 86%, respectively, and fall within the range of published values for *B. latifrons* (McDonald, 1981).

Horn cores. The right horn core is largely intact, while the ventral 80% of the left horn core was scraped away by grading equipment during construction. The base of the right horn core has been displaced anteriorly such that the ventral portion of its anterior margin now aligns with the posterior rim of the orbit. More distally, the right horn core displays a distinctly concave posterior margin as viewed dorsally. The distal tip is intact and terminates at a point posterior to the nuchal crest and slightly elevated above the level of the frontals forming the skull roof. Vascular grooves on the horn core suggest straight growth rather than spiral growth along the longitudinal axis. There is no obvious bony ridge or burr at the base of either horn core, a condition generally associated with female or immature male bison (McDonald, 1981). The absence, however, of these structures may be a preservational artifact rather than an actual developmental feature of the specimen. Overall, the intact right horn core does not display the level of distal tapering preserved in published specimens of adult *B. latifrons*. Instead, the horn core remains relatively robust nearly to its distal termination (Figure 4A).

Jugal. The right jugal is relatively short anteroposteriorly in dorsal aspect and deep dorsoventrally when compared with the more gracile jugal of *Bos taurus* and *Bison bison*. In ventral aspect the right jugal is well exposed and expanded laterally, with a sharply defined lateral margin marking the lateral extent of the origin of the masseter muscle. The temporal process of the jugal is also transversely broad where it forms the ventral margin of the laterally protruding orbit.

Nasals. The nasals are undeformed for most of their length and transversely very broad, measuring 103.6 mm across at the level of the narial incision. Anteriorly, they are bluntly triangular and overhang the external nares to form a distinct narial incision with the adjacent premaxilla. Unlike the condition in most species of long horned bison, the dorsal surface formed by the right and left nasals is low and relatively flattened. The suture with the adjacent maxilla is horizontal and lies essentially at the lateral edge of the snout rather than being medially retracted. This configuration correlates with the steeper slope of the lateral surface of the maxilla.

Palatine. The right and left palatines are broken and deformed especially posteriorly where the internal narial passage has been obscured by matrix and displaced bony elements. The medial portion of the right palatine is better preserved than the left and is marked by a large diameter medial palatine foramen positioned immediately posterior to the damaged maxilla-palatine suture.

Squamosal. Portions of the left squamosal were exposed and include a partial glenoid fossa and the finger-like zygomatic process. These elements have been displaced anteriorly as part of the deformation of the occipital and basicranial region. The right squamosal has been obscured by crushing of the right paroccipital process.

Table 2. *Measurements of upper dentition for SDSNH 135465. AP= maximum anteroposterior crown diameter; TR= maximum transverse crown diameter*

Tooth position	AP (mm)	TR (mm)
LP3		20.0
LP4	18.9	22
RP4	19.7	22.3
RM1	31.1	26.7
RM2	33.3	27.3
RM3	34.1	25.4

Basicranium. The basioccipital has been rotated anterodorsally and displaced to the right. This distortion has obscured the region of the internal narial opening, as well as the right glenoid fossa and much of the left glenoid fossa. The basioccipital crests are relatively robust and protrude ventrolaterally. The left auditory bulla was exposed after removal of the displaced left postglenoid process of the squamosal and closely approaches the general bulbous shape of other bovid bullae.

Dentition. Measurements of the upper dentition are summarized in Table 2. The left P3-4 and right P4 are preserved in their respective maxillae (Figure 5). The left P3 has been rotated posteriorly such that the surface of the occlusal surface of the crown is lying against the anterior margin of the P4 crown. In this configuration the crown morphology of P3 is not visible. The crowns of the other premolars, however, are well exposed and reflect a degree of crown wear consistent with a mature adult individual. The right M1-3 series is preserved in the right maxilla, which is distorted to the point that the M3 has been dorsally displaced relative to M1-2. Like the premolars the degree of crown wear is consistent with a mature adult individual. Overall, the premolars and molars of SDSNH 135465 are relatively small, a condition noted earlier by Vanderhoof (1942) who commented that in spite of its large body size the cheek teeth of *B. latifrons* are about the same size as those of modern *B. bison*. McDonald (1981) utilized only two dental morphometric characters:

M1-3 labial length, and M3 anterior cups transverse width. For both measurements, the teeth of SDSNH 135465 are about 15% smaller than published records of male and female *B. latifrons*.

Postcrania. Recovered postcranial elements include an isolated axis vertebra, an associated series of three posterior cervical vertebrae and one anterior thoracic vertebra, a semi-articulated series of three posterior thoracic and five lumbar vertebrae, a fused series of five sacral vertebrae, the right and left innominates, a right humerus, and a fused right radius-ulna. The cervical vertebrae were found in close proximity and are presumably from the same individual. Because they were collected approximately 52 meters from the skull, they have been catalogued separately as SDSNH 135958. Both transverse processes are preserved on L2-L5, while L1 has only the left transverse process. The neural spines and neural arches were scraped away from all lumbar and thoracic vertebrae. The sacral vertebrae are tightly ankylosed into a single sacral element measuring 244 mm in anteroposterior length. The sacrum, in turn, is partially articulated with the right and left innominates. The sacroiliac joint is open and the sacral vertebrae have shifted ventrally relative to the pelvic bones. The innominates are complete and strongly fused at the pubic symphysis. Internally, the three bones making up each innominate (ilium, ischium, and pubis) are also strongly fused, suggesting that the individual was a mature individual. The right humerus (Figure 6A, B) has a functional length (humeral head to distal condyle) of 355 mm, an anteroposterior diaphysis minimum diameter of 71 mm, and a transverse minimum diphysis diameter of 66 mm. Except for some equipment damage to the lateral epicondyle, the humerus is complete. The medial portion of the robust lateral tuberosity only slightly overhangs the bicippital groove, a feature characteristic of *Bison* and distinct from the broadly overhanging lateral tuberosity in species of *Bos*. The deltoid tuberosity on the lateral border of the diaphysis is developed as an elongate raised process as in *Bison*. The right radius and ulna (Figure 6C, D) are tightly fused together along the radial diaphysis.

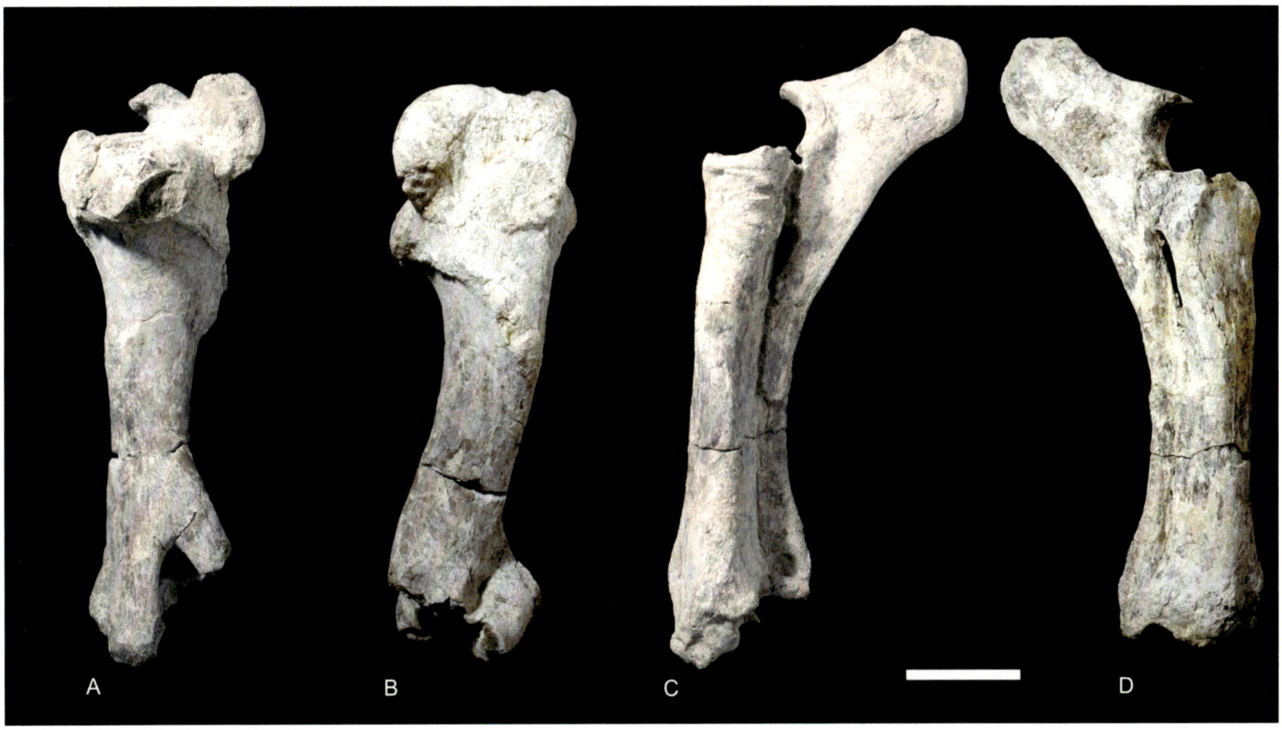

Figure 6. Bison latifrons *(SDSNH135466) forelimb bones. A, right humerus (caudal view); B, right humerus (cranial view); C, right radius-ulna (medial view); D, right radius-ulna (lateral view). Scale bar equals 10 cm.*

Both are complete except for equipment damage to the radial tuberosity. The radius has a functional length (capitular fossa to distal condyle) of 356 mm, while the total length of the ulna (olecranon process to distal styloid process) measures 459 mm. The transverse minimum diaphysis diameter of the radius is 62.5 mm. The humerus and radius-ulna were found in close proximity and are presumably from the same individual. Because they were collected approximately 40 meters from the skull, they have been catalogued separately as SDSNH 135466.

DISCUSSION

Taxonomic identifications of species of North American *Bison* have been primarily based on morphometric features of the horn cores. This is a bit surprising given the obvious social display function of such structures and the apparent morphological complexity of other parts of the skull, dentition, and postcrania. However, numerous authors have pointed out the phylogenetically conservative aspects of bovid cranial and postcranial anatomy and the difficulty presented by the generalized nature of many anatomical features (McDonald, 1981 and references therein). Having said this, certain morphometric features of the Pala Mesa bison fossils can be used to assign these specimens to the extinct giant bison, *Bison latifrons*. These features include the large size of the horn cores, the lack of longitudinal spiraling of the distal portion of the horn cores, the lack of doming of the frontals, the high degree of relative orbital protrusion, and the overall large size of all recovered skeletal elements. A review of the morphometric conclusions of McDonald (1981) concerning sexually dimorphic features of the skull of *B. latifrons*, suggests that the Pala Mesa bison likely is a female individual. These features include the relatively shorter length of the horn cores, the presence of a slight dorsal burr at the base of the horn cores, and the relatively short skull length between the nuchal crest and the tip of the premaxillae. Although some of these features have also been reported in immature male specimens of *B. latifrons*, the fused condition of most cranial sutures and the degree of wear on the occlusal surfaces of cheek tooth crowns suggests that SDSNH 135465 was a mature adult individual. Thus, it appears that the Pala Mesa bison represents an adult female specimen of *Bison latifrons*.

The earliest occurrence of *Bison* in North America represents an important biochronological datum that marks the initial dispersal of this lineage from Eurasia over the Bering land bridge and into the New World. This datum has been employed by vertebrate paleontologists to mark the beginning of the Rancholabrean North American Land Mammal Age, which is roughly calibrated at between 210± 60 ka and about 160 ka (Bell et al., 2004; Scott and Cox, 2008). This date represents a maximum age for the Pleistocene strata exposed at the State Route 76–Interstate 15 interchange. McDonald (1981) further proposed that the maximum population numbers and geographic distribution of *Bison latifrons* was during the Sangamon interglacial approximately 120 ka and that extinction of this species occurred during the late Wisconsin glaciation between approximately 20–30 ka. Future geochronologic studies of SDSNH 135465 will focus on obtaining accurate radiometric dates for this site.

SUMMARY

Paleontological mitigation work during construction of the State Route 76—Interstate 15 interchange in the San Luis Rey River Valley near Pala Mesa resulted in the discovery of well-preserved fossil remains of *Bison*. Morphometric analysis of these fossil remains, especially a nearly complete skull (SDSNH 135465), suggest that the fossils belong to the extinct giant bison, *Bison latifrons*. Cranial suture fusion, occlusal wear of cheek teeth, and sexual dimorphic features suggest the specimens may represent a mature, female individual. The Pala Mesa specimen of *Bison latifrons* represents the first record of this species in the Pleistocene of San Diego County.

ACKNOWLEDGEMENTS

The authors wish to acknowledge Caltrans, District 11 for supporting mitigation of potential negative impacts to paleontological resources within its jurisdiction. Special thanks go to Laurie Berman, District 11 Director; Bruce April, District 11 Deputy District Director; Kevin Hovey, District 11 Environmental Analysis Branch Chief; and Karen Crafts, District 11 Associate Environmental Planner. The bison fossils were discovered and recovered by SDNHM paleontologists including Nicole Anderson, Gino Calvano, and Brad Riney. Preparation of the bison fossils was done in the SDNHM Department of Paleontology by Nicole Anderson and Merry Whittington. Graphics were prepared by Shelly Donohue, Department of Paleontology, SDNHM.

REFERENCES CITED

Bell, C.J., E.L. Lundelius, Jr., A.D. Barnosky, R.W. Graham, E.H. Lindsay, D.R. Ruez, Jr., H.A. Semken, Jr., S.D. Webb, and R.J. Zakrzewski. 2004. The Blancan, Irvingtonian, and Rancholabrean Mammal Ages. In, M.O. Woodburne (ed.), Late Cretaceous and Cenozoic mammals of North America: Biostratigraphy and geochronology. New York: Columbia University Press. 232–314.

California Department of Transportation. 2012. *Caltrans Standard Environmental Reference, Environmental Handbook.* Volume 1, Chapter 8, Paleontology. Available: http://www.dot.ca.gov/ser/vol1/sec3/physical/Ch08Paleo/chap08paleo.htm#pmp. Accessed: June 4, 2013.

Guthrie, D.A., 2012. Avian material from Rancho del Oro, a Pleistocene locality in San Diego County, California. Bulletin of the Southern California Academy of Sciences: 109:1–7.

Jefferson, G.T. 1991. A catalogue of late Quaternary vertebrates from California: Part two, mammals. Natural History Museum of Los Angeles County, Technical Reports 7:1–129 (and unpublished addendum).

Kennedy, M.P., and S.S. Tan, 2005. Geologic Map of the Oceanside 30' x 60' Quadrangle, California: a Digital Database. California Geological Survey, Preliminary Geologic Maps, scale 1:100,000.

Kern, J.P. and T.K. Rockwell. 1992. Chronology and deformation of Quaternary marine shorelines, San Diego County, California. In, Quaternary Coasts of the United States: Marine and Lacustrine Systems. Society of Economic Paleontologists and Mineralogists, Special Publication 48:377–382.

Majors , C.P. 1993. Preliminary report on a late Pleistocene vertebrate assemblage from Bonita, San Diego County, California. PaleoBios 15:63–77.

McDonald, J.N., 1981. North American bison: their classification and evolution. University of California Press, Berkeley, 316 pp.

McDonald, J. N., and G.E. Lammers, 2002. *Bison antiquus* from Kenora, Ontario, and Notes on the Evolution of North American Holocene *Bison*. In, R.J. Emry (ed.), Cenozoic mammals of land and sea: tributes to the career of Clayton E. Ray. Smithsonian Contributions to Paleobiology No. 93, Washington, D.C., pp. 83–97.

Miller, W.E. 1968. Occurrence of a giant bison, *Bison latifrons*, and a slender-limbed camel, *Tanupolama*, at Rancho La Brea. Los Angeles County Museum, Contributions in Science 128:1–9.

Scott, E. and S.M. Cox. 2008. Late Pleistocene distribution of *Bison* (Mammalia; Artiodactyla) in the Mojave Desert of Southern California and Nevada. In, X. Wang and L.G. Barnes (eds.), Geology and Vertebrate paleontology of Western and Southern North America. Natural Histroy Museum of Los Angeles County, Science Series 41: 359–382.

Skinner, M.F. and O.C. Kaisen. 1947. The fossil Bison of Alaska and preliminary revision of the genus. Bulletin, American Museum of Natural History 89:123–256.

Tan, S.S., 2000. Geologic map of the Bonsall 7.5-minute quadrangle, San Diego County, California: A digital database. California Geological Survey, Preliminary Geologic Maps, scale 1:24000.

Tan, S.S. and M.P. Kennedy, 1996. Geologic Maps of the Northwestern Part of San Diego County, Plate 1, Geologic Map of Oceanside, San Luis Rey, and San Marcos 7.5' Quadrangles, San Diego, California. California Division of Mines and Geology, Open-file report 96-02, scale 1:24,000.

Vanderhoof, V.L. 1942. A skull of *Bison latifrons* from the Pleistocene of northern California, University of California Publications, Bulletin of the Department of Geological Sciences 27 (1): 1–23.

Walawender, M.J., 2000. The Peninsular Ranges: A Geological Guide to San Diego's Back Country. Kendall/Hunt Publishing Company, Dubuque, Iowa.

Medicine Rock on the Pala Indian Reservation. Author Norrie Robbins for scale.

The San Luis Rey Watershed as Native Land

Eleanora I. Robbins, PhD
San Diego State University
Department of Geological Sciences
norrierobbins@cox.net
619-303-9095

Shasta C. Gaughen, PhD
Director, Pala Environmental Department
Christopher A. Ortiz
Los Coyotes Environmental Department
Andrew R. Pigniolo
Laguna Mountain Environmental, Inc.

The watershed of the San Luis Rey River (SLR) (Figure 1) is the traditional territory of the Luiseño and Cupeño people. The Luiseño have lived along this river (Waníicha) for at least 2,000 years, travelling annually up Palomar Mountain (Pa'áa'aw) to harvest acorns. The archaeological record shows a long history of human occupancy in the region. D.L. True, an archaeologist who grew up in the Pauma Valley, identified a pattern of early-age sites he labeled the Pauma Complex (True 1958). Viewed today, these sites probably represent a mixture of Paleoarchaic large bifaces and cresentics and Archaic period milling technology and stone tools. The Late Prehistoric archaeological record has been divided into periods based on the use of ceramics.

The archaeology within the watershed shows a high degree of human adaptation to local resources, from the intensive use of acorns and associated bedrock milling in the upper reaches, to the early use of estuary shellfish in the SLR embayment, shifting to the intensive late use of sandy beach shellfish (*Donax*) in the coastal zone. Lithic material usage also shows a high degree of adaptation from poor quality aplites and quarzites in the Los Coyotes area, to the use of pebble quartz from the San Onofre Breccia and Santiago Peak Volcanics and Piedra de Lumbre chert from outside the immediate watershed. The use of quartz dominates the lithic pattern in the San Luis Rey watershed (True 1966).

At the time of European arrival in San Diego in 1769, there were well over 50 village, encampment, and resource processing sites. With the arrival of Europeans, the people were either incorporated into the San Luis Rey or the San Antonio de Pala Missions or else they fled into the mountains. Due to disease, deprivation, and the hardships of life in the mission system, the native population plummeted from several thousand to several hundred by the time California became a state in 1851 (Carrico 2008).

Ultimately the Federal government divided the Luiseño and Cupeño into four bands on separate reservations which were created by executive order in 1875: Pala, Pauma, Rincon, and La Jolla. The San Luis Rey band, associated with the San Luis Rey Mission in Oceanside, was never granted federal status. Also in the 1870s, Los Coyotes (Cahuilla and Cupeño people) and San Pasqual (Kumeyaay people) Reservations were created in the watershed. Most Cupeño were moved forcibly from Warner Springs to Pala (Carrico 2008).

Access to water was a critical factor in determining the location of permanent villages. At least a dozen villages lined the SLR and its tributaries. Water was also critical for processing acorns, the major staple food source. The presence of large granitic outcrops associated with oak groves and even minor waterways generally indicates an increased likelihood of archaeological sites and artifacts. Water is a prominent feature in Luiseño and Cupeño histories of the world's beginnings. Drought, if it occurred, does not seem to have been significant enough to be mentioned in oral histories. In fact, the river was seen to be perennial in recorded history. Luiseño elders can recall their own parents and grandparents washing clothes in

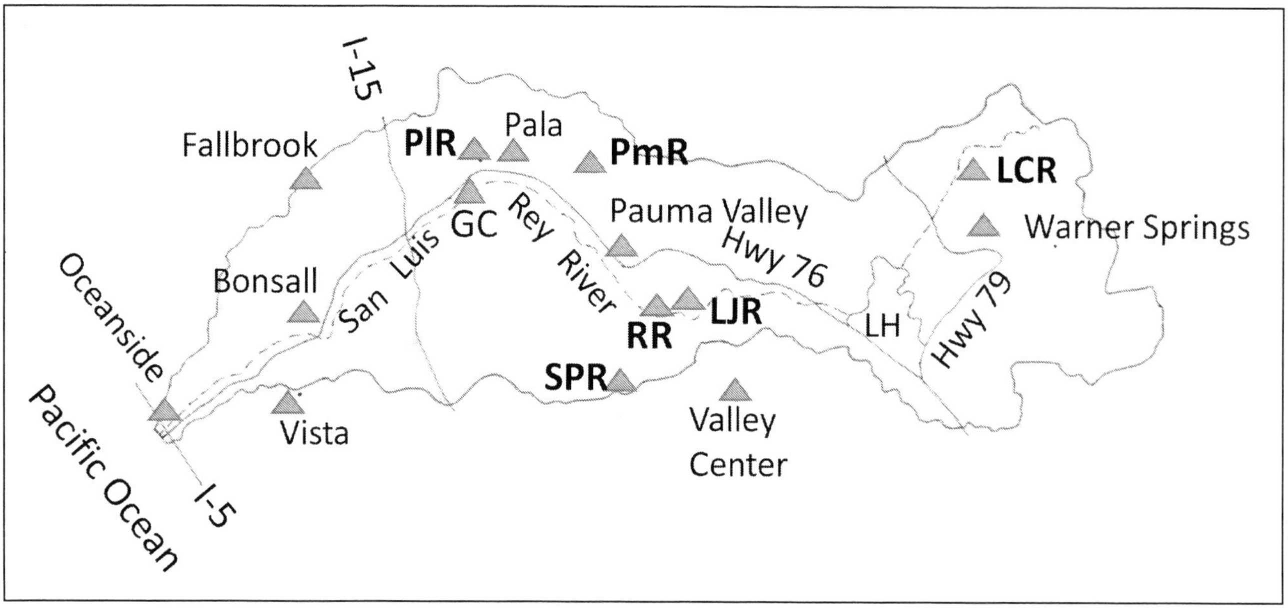

Figure 1. *San Luis Rey River watershed. (Abbreviations: GC-Gregory Canyon, LCR-Los Coyotes Reservation, LH-Lake Henshaw, LJR-La Jolla Reservation, PmR-Pauma Reservation, PIR-Pala Reservation, RR-Rincon Reservation, SPR-San Pasqual Reservation)*

the river, catching fish, and swimming. Acorns and other vegetable foods such as cactus fruits were processed on the river banks. Tule boats made of reeds were also used for fishing during the mission era and in precontact times. Those who grew up on the Pala reservation during the 1960s and 1970s have stories of swimming in the river and jumping into deep water from the "swinging bridge," which was the only access over the river when it was in flood and impassable by automobiles (Gaughen 2011). The swinging bridge is still standing just east of the Lilac Road overcrossing. Living close to the Elsinore fault, the people discussed earthquakes both in their oral history and songs.

To support the burgeoning Euro-American population, SLR River was dammed to form Lake Henshaw in 1922. Water ceased to flow during the dry season, which was a recognized hardship for all who lived along the river (San Luis Rey Indian Water Authority, n.d.[a]). La Jolla Tribal Elder Henry Rodriguez, who was then Chair of the SLR Indian Water Authority testified to the Senate in 1976 about the effects of water scarcity on his Tribe and the other Bands with water rights to the San Luis Rey, which included La Jolla, Rincon, San Pasqual, Pauma, and Pala. Ultimately, U.S. Congress passed the San Luis Rey Indian Water Rights Settlement Acts of 1988 and 2000 (Public Law 100-675 in 1988, and Public Law 106-377 in 2000), which required that Lake Henshaw release 17% of any water conserved, up to 16,000 acre-feet, to the portion of SLR River that moved through the La Jolla reservation so that it could be distributed amongst the five Bands. Once water leaves the La Jolla reservation, it is diverted into the Escondido Canal which carries water to Vista and Escondido. Although the five bands' water rights were restored by the two Acts, a final settlement of how the water will be distributed has yet to be reached and no additional water has been released to the SLR (San Luis Rey Indian Water Authority, n.d.[b]). Today, the upper watershed has remained relatively undeveloped, whereas the middle watershed is primarily devoted to agriculture, and the lower watershed is urban (Figure 1). The water chemistry (Table 1) reflects this tripartite nature. Table 1, which compares winter and spring chemistry data sampled at times when there was water in the river. The upper watershed site is a mountain-derived 'freshet,' having values that could be interpreted as the naturally-occurring water chemistry previous to European colonization. The high arsenic values probably reflect naturally-occurring arsenic in the Julian Schist that crops out only at the very top of the watershed. Downstream increases in chloride probably reflect input of treated water; increases in sulfate probably reflect input of high sulfate Colorado River water, which is a major water source in many urban environments in the county. Agriculture probably accounts for downstream increases in phosphate. Total dissolved solids (TDS) in shallow groundwater in the San Luis Rey River Valley range from below 250 milligrams per liter near the crest of the watershed to as much as 2,700 mg/L in coastal Oceanside (PBS&J 2003).

Table 1. *Water chemistry data for winter and spring at three sites along San Luis Rey River.*

	Watershed location	**Lower**	**Lower**	**Middle**	**Middle**	**Upper**	**Upper**
	Collecting Locality	**Oceanside SLR-8**	**Oceanside SLR-8**	**Above Greg. Cyn. SW-1**	**Above Greg. Cyn. SW-1**	**Los Coyotes**	**Los Coyotes**
	Season	**Winter**	**Spring**	**Winter**	**Spring**	**Winter**	**Spring**
	Date	**10/27/2011**	**4/15/2005**	**12/-/2000**	**3/-/2001**	**2/19/2012**	**4/7/2012**
Analyte	**Units**						
Temp	°C		18.47			13.1	13.1
TDS	mg/L			657	627	244	246
Turbidity	NTU		14.3			0.26	0.68
SC	uS/cm		2052			382	377
pH	units		7.57	7.92	8.18	6.87	6.64
NO3-N	mg/L					2.4	2.2
NO3	mg/L			0.03	0.03		
NH3-N	mg/L		0.12			0.2	0.26
PO4+P	mg/L	397.4	0.18			0.16	0.15
SO4	mg/L		487	185	183	34	74
Ca	mg/L			81	98	29	30
Mg	mg/L			30	34	11	11
Na	mg/L			67	76	24	26
K	mg/L					2.1	2.4
Sb	ug/L			10	8	71	26
As	ug/L	2.45	1.13	2	1	<3	34
Cd	ug/L	0.15	0.05	4	4	<2	<2
Cl	mg/L			103	107	28.2	26.8
Cr	ug/L	34.6	0.51	5	4	27	12
Cu	mg/L	19.1	5.06	0.01	0	<10	<10
Fe	mg/L					<0.052	<0.052
Pb	ug/L	10.4	0.07	4	0.4	<2	2.2
Mn	mg/L	798	3.08			<0.01	<0.01
Se	ug/L		16	0.6	0.6	<6.5	<6.5
Th	ug/L			20	0.7	<2	6.8
V	ug/L			10	3	<7	7.4
Zn	ug/L			6.6	2	0.27	0.11
	Reference	SWRCB unpub. data	SWRCB unpub. data	SWRCB, 2003	SWRCB, 2003	Sierra Anal., 2012	Sierra Anal., 2012

THE LITHOLOGICAL TOOLBOX

Through time, the people had and still have a sophisticated knowledge and uses of rocks and minerals in their territory. Large granitic boulders were used to form bedrock mortars. Pestles and stand-alone mortars were fashioned out of granitic and rare andesitic cobbles or rare sandstone boulders (True et al. 1974). Large flat surfaces in boulder piles formed surfaces for petroglyphs, or rock paintings; many such boulder configurations became ritual sites. The red pigment of the paintings was created from the flocculates of iron seeps (Campbell 2007) or the weathered rims of iron-bearing rocks. Black pigment was created from charcoal augmented by oxidized manganese (Campbell 2007); manganese-stained cobbles are present in several creeks along the watershed (Robbins, pers. obs.). Living along the Pala Mineral Belt, the Luiseño people recognized tourmaline and used it ceremonially (True et al. 1974). Tourmalines and quartz crystals were important components of the ceremonial bundles used by the shaman, or *Not^{h}*, and were often included as part of the grave goods that were included in traditional inhumations and cremations in pre-contact times (Boscana 1933; Kroeber 1925; Strong 1929; Sparkman 1908).

An analysis of Late Paleolithic (older than 1000 years B.P.) sites in the watershed by Dietler (2004) revealed that the toolbox of the people included rocks and clay from nearby and far sources (Table 2). Most of the rocks used for tools were available in the watershed. Santiago Peak Volcanics supplied aphanitic andesite, pyroclastic deposits, quartz latite, rhyolite, and minor basalt (Stephens 2009). La Jolla-Poway Group conglomerates supplied dacite, rhyolite, and quartzite (Abeid 2010). Both were readily available in the western part of the watershed. Useable quartz comes from pegmatite dikes in the Peninsular Range batholith, and especially around Pala. Cryptocrystalline quartz, primarily Piedra de Lumbre and Talega Canyon cherts were quarried from Camp Pendleton (Robbins et al. 2012). Obsidian does not occur anywhere in San Diego County, but it is so important that it was traded and valuable enough to carry for long distances; the obsidian found in watershed sites is from Obsidian Butte at the southeast edge of Salton Sea and from Coso Hot Springs in Owens Valley (Pigniolo 1996). Silicified Bedford Canyon Formation and Julian Schist metasediments crop out in the watershed. Steatite for implements typically is from Santa Catalina Island, but lacking chemical data, other sources cannot be ruled out (Stonewall Peak in Cuyamaca Rancho State Park, Jacumba Valley, Boiling Springs on Mount Laguna, and quarries in the Rincon and Valley Center areas) (Graham 1981; Heizer and Treganza 1944; O'Neil 1983; Parkman 1983; Polk 1972; Treganza 1942).

Iron-rich clay played an important role in pottery making. The pottery type made in the watershed is called Palomar Brown or Tizon Brown Ware (Meighan 1959; Sutton 2011). The pottery

Table 2. *Percentages of lithological materials from 20 archaeological sites in the San Luis Rey watershed. (From Dietler, 2004)*

Lithology	Western zone	Central zone	Eastern zone	Far Eastern zone
Volcanics	62.97	37.01	27.20	0.96
Quartz	19.95	50.46	56.45	93.83
Cryptocrystalline quartz (chert)	12.83	9.38	9.60	0.54
Quartzite	2.36	1.12	0.00	2.09
Obsidian	0.77	1.54	3.20	2.14
Bedford Canyon metasediments	0.00	0.00	0.00	0.46
Other (includes steatite)	1.12	0.47	3.55	0.00

Table 3. *Lithology of worked materials at the Molpa Archaeological Site (True et al., 1974; P. Davis, written commun.; Robbins, pers. obs., 2013)*

Points	67% quartz (23% clear, minor jasper), 15% chert 7% obsidian 6% microbreccia, minor metasediment (gray to dark gray) 1% Santiago Peak Volcanics
Crystals	quartz (clear), tourmaline (black)
Scrapers	metasediment, basalt, quartz, jasper
Arrow-shaft straightener	steatite (dark gray)
Clay objects	pipes (dark gray, reddish brown), beads, pottery
Cooking/boiling stone	basalt

also contains mica, quartz, and feldspar; the red-weathered gabbro around the La Jolla Reservation would have been a good source for such clay (Robbins, pers. obs.). A clay deposit has been mapped along the slopes of Palomar Mountain (Heizer and Treganza 1944); it may or may not be the locus of early-worked clay, but many outcrops of red-weathered gabbro are mapped west of the mountain.

Detailed lithologic analysis of one ancient village in the watershed. The Molpa archaeological site is situated 0.3 mile south of California State Route 76 and one mile east of the South Grade turnoff to Palomar Mountain. This ancient Luiseño settlement encompasses 3-5 acres in hilly terrain having scattered granitic outcrops. Artifacts are concentrated mainly in two areas; a westerly midden site excavated by UCLA (designated SDi308) in 1955–1957 (True et al. 1974), and an area about 300 feet to the east of the UCLA dig, which is the source of points described below. The Molpa Site was occupied for several thousand years and was abandoned in the early 1800s. It was a Luiseño winter village, at elevation 2650 feet; summer villages were sites predominantly on Palomar Mountain.

A total of 189 projectile points, either whole or at least half of their original length, were excavated and screened (P. Davis, written commun. 2013). The whole, unbroken points ranged from 0.5 to 1.88 inches in length, and averaged just under one inch long. Projectile point statistics are shown on Table 3, along with the full range of lithologies found at the site. The projectile point percentages are comparable to those found on Palomar Mountain at a summer campsite (SDi-593) where projectile points were 79% quartz, 12% chert, 3% obsidian, and 2% basalt. Several chert and obsidian points from the Molpa site were perfectly symmetrical, serrated, and extremely well crafted.

PROPOSED LANDFILL IN GREGORY CANYON

An environmental issue that will affect all who live west of the Pala Reservation is a proposed landfill that would be sited in Gregory Canyon, along the banks of the SLR (Figure 1). In the late 1980s, when the site was proposed for a solid waste landfill, it was rejected by San Diego County because the site did not meet the several criteria necessary for landfill construction. The private developers then ran a local ballot initiative in 1994 to change the zoning for the property and circumvent the County's land use and planning procedures.

Located on the western border of the Pala Reservation, Gregory Mountain forms a significant feature on the landscape both geologically and culturally. Known as *Chokla* or *Takwic Puki* in the Luiseño language, Gregory Mountain is one of the acknowledged homes of the spirit Takwic. Takwic is one of the First People, and he holds an important place in Luiseño cosmology (Harrington 1933; Strong 1929). According to oral history, Takwic gathers the souls of the dead. Stories tell of his dwelling on Chokla, either in a cave or on a cliff overlooking Pala. Takwic travels through the sky as a ball of fire or streak of light and is associated with lightning. If one should happen to see Takwic in the sky, it is best to look away so as not to attract the spirit's attention (Bean 1992, 1995).

Since the passage of the 1994 initiative, the Pala Band of Mission Indians and several community and environmental groups have fought to prevent the landfill. For Pala, the idea of siting a landfill on the flanks of sacred Chokla, the home of Takwic, is an enormous threat and desecration of traditional culture. The entire area surrounding Gregory Mountain is environmentally and culturally sensitive. Medicine Rock, a large granitic boulder standing in isolation from other rock formations, is decorated with petroglyphs and is a site of extreme significance to Luiseños. Currently used as a site of prayer, meditation, healing, and contemplation, Medicine Rock was used in earlier times as an integral part of adolescent puberty rites that took place throughout the adjacent river valley and canyon (Dubois 1908; Kroeber 1925; Sparkman 1908; Boscana 1933; Harrington 1933). Numerous archaeological sites, from isolated bedrock mortars to complex settlement areas, surround the base of Gregory Mountain and line the river.

Not only is the site of cultural and religious significance to the people, it is also poorly chosen because of seismicity issues, fire frequency, and pollution of a major source of water to the downstream cities of Bonsall and Oceanside.

The potential landfill site is within 4 km of the Elsinore-Whittier fault, the most active fault in southern California. A small search of the California Institute of Technology database of earthquakes of Magnitude 3 or greater felt within 30 miles of Pala, California, showed 35 earthquakes between January 1, 2002 and June 30, 2012 (California Inst. Tech., Southern California Earthquake Data Center, *http://www.data.scec.org*). This highlights the frequency of pounding and shaking in the vicinity of the landfill site, consistent with the results of Madabhushi and Singh (2001) who studied liner breaching associated with seismicity. Although the developers are promoting the proposed landfill liner as the most technologically advanced and environmentally protective in the nation, the U.S. EPA concurs that, over time, all landfill liners eventually break down and allow leachate contamination (U.S. Environmental Protection Agency 1995).

Methane discharge from a landfill is another serious issue for a watershed that has been subject to major fires during the hot summer months. The north-facing slope of Gregory Mountain is known to be an area subjected to an inordinate number of lightning strikes, thereby earning its reputation as the home of Takwic; the mining engineer who owns two working tourmaline mines on Palomar Mountain across the valley related that he "likes to sit outside and watch the lightning hit Gregory Mountain in the summer" (J.B. Shepard, oral commun. 2003). Lightning is a known problem with methane-generating landfills in general (Montague 1990). Lastly, the SLR flows past the mouth of Gregory Canyon, making it extremely vulnerable to pollution from toxic landfill leachate. Gregory Canyon itself is bisected by an intermittent stream that would be completely filled in by landfill construction. Pala and the communities to the west of Gregory Mountain draw much of their drinking water from the SLR and from the fractured bedrock aquifer that is recharged by seasonal rains and runoff. A landfill in Gregory Canyon would effectively channel polluted runoff towards the river and into the aquifer either through surface flows during rain events, or through the leaching of pollutants through the ground.

These are all significant concerns, which is why the proposed landfill along Gregory Mountain is so contentious. There are other concerns that are addressed in the Draft Environmental Impact Statement (County of San Diego 2012).

CONCLUSIONS

For as long as 10,000 years, the watershed has supplied the critical resources necessary to support large populations of Native Americans. The number and complexity of the archaeological sites are testaments to the importance of the San Luis Rey River in the pre-contact era. The rich ecology, geography, and geology of the SLR watershed supported a diversity of complex Native American cultural traditions. In historic times, the river has allowed the development of intensive agriculture throughout the watershed, and has also supported the growth of urban areas by providing a municipal water source.

ACKNOWLEDGMENTS

We thank Paul Davis for access to Molpa site artifacts; Eric Elliott for Luiseño translations; Penni Rubin for drafting aid; Lilian Busse (California State Water Resources Control Board), Joanna Wisniewska (County of San Diego, Department of Public Works), and Suzanne Crowley (U.S. Geological Survey) for water-quality data retrieval.

REFERENCES

Abeid, John, 2010, Late Cretaceous-Eocene landscape evolution and drainage reorganization along the southwest edge of the Cordillera; insight from volcanic clasts in conglomerates of the Cabrillo Formation and Poway-La Jolla Groups, San Diego County, California: Geol. Soc. Amer., Abs. with Programs, v. 42, p. 75.

Bean, Lowell J., 1992, Power and Its Applications in Native California in California Indian Shamanism, Ballena Press, Menlo Park, California, p. 22–25.

Bean, L.J, 1995, Takwish, in L.J. Bean, Jerry Schaefer, and Sylvia Brakke (eds.), Archaeological, Ethnographic, and Ethnohistoric Investigations at Tahquitz Canyon, Palm Springs, California: Menlo Park, CA, Cultural Systems Research, pp. F1–F8.

Bean, L.J. and Shipek, F.C., 1978, Luiseño, in R.F. Heizer (ed.), Handbook of North American Indians, Volume 8, California: Smithsonian Institution, Washington DC, p. 550–563.

Boscana, Geronimo, 1933, Chinigchinish, in P.T. Hanna (ed.), Chinigchinish: Fine Arts Press, Santa Ana, California, p. 17–90.

Brian A. Stirrat & Associates (BAS). 2003. Joint Technical Document for the Gregory Canyon Landfill. http://www.waterboards.ca.gov. Appendix G tables retrieved September 16, 2013, from http://www.waterboards.ca.gov/rwqcb9/water_issues/programs/land_discharge/docs/canyon_project/jtd_volume_ii_a/appendix_g/appndx_gtables.pdf

Campbell, P.D., 2007, Earth Pigments and Paint of the California Indians: Los Angeles, CA, Campbell Pubs., p. 73.

Carrico, Richard, 2008, Strangers in a Stolen Land: Sunbelt Pubs., San Diego, CA, 203 p.

County of San Diego, 2012, Gregory Canyon landfill. www.sdcounty.ca.gov. Accessed September 16, 2013, from www.sdcounty.ca.gov/deh/waste/chd_gregory.html

Dietler, John, 2004, Lithic material use in Late Prehistoric San Diego County: Proc. Soc. Calif. Arch., v. 14, p. 57–67.

DuBois, C.G., 1908, The Religion of the Luiseño Indians of Southern California: Univ. Calif. Publ. in Amer. Arch. and Ethnology, v. 8, no. 3, Berkeley, CA, The University Press, 186 p..

Gaughen, S.C., 2011, Against the odds: Indian gaming, political economy, and identity on the Pala Indian reservation: PhD Dissertation, Univ. New Mexico, 256 p.

Graham, W.R., 1981, Final Report Cultural Resource Survey of the Laguna Mountain Recreation Area, San Diego County, California: Archaeological Systems Management, Inc.

Harrington, J.P., 1933, The Papers of John Peabody Harrington in the Smithsonian Institution, Volume 3: Microfilm Reel 119, Southern California Basin.

Heizer, R.F., and Treganza, A.E., 1944, Mines and quarries of the Indians of California: Calif. J. Mines and Geology, v. 40, p. 285–359.

Kroeber, A.L, 1925 [1976], Handbook of the Indians of California: New York, Dover Pubs, 995 p.

Madabhushi, S.P.G., and Singh, S., 2001, Behaviour of Landfill Liners under Earthquake Loading, in B.H.V. Topping (ed.), Proceedings of the Eighth International Conference on Civil and Structural Engineering Computing, Civil-Comp Press, Stirlingshire, UK, Paper 122, 2001.

Meighan, C.W., 1959, Archaeological resources of Borrego State Park: Univ. Calif. Arch. Survey Ann. Rpt., 1958–1959, Los Angeles, p. 36–38.

Montague, P. 1990. Unrecognized Cause of Landfill Failures. Rachel's Hazardous Waste News No. 177. April 18. Retrieved September 16, 2013, from http://www.ejnet.

org/rachel/rhwn177.htm

O'Neil, D.H., 1983, A Shaman's "sucking tube" from San Diego County, California: J. California and Great Basin Anthro., v. 5, p. 245–247.

Parkman, E.B., 1983, Soapstone for the Cosmos: Archaeological Discoveries in the Cuyamaca Mountains: J. California and Great Basin Anthro., v. 5, p. 140–155.

PBS&J, 2003, San Luis Rey River Watershed Urban Runoff Management Program (WURMP), Prepared for California Regional Water Quality Control Board, San Diego Region. January, p. 3-27 and 3-28. Retrieved September 15, 2013, from www.sandiegohealth.org/water/runoff/slr_river_wurmp_final.pdf

Pigniolo, A.R., 1996, Lithic material types as a chronological indicator in the archaeological record of San Diego County: Proc. Soc. Calif. Archaeology, v. 9, p. 209–220.

Polk, M.R., 1972, Manufacture and use of steatite objects by the Diegueño: Pacific Coast Arch. Soc. Quart., v. 8, p. 5–26.

Pub. L. No. 106-377, Appendix B, 114 Stat 1441 A (2000). Print.

Robbins, E.I., Pigniolo, A.P., Cranham, G.T., and Elliott, W.J., 2012, Preliminary analysis of Piedra de Lumbre and Talega Canyon cherts: Distinctive and historically significant outcrops on Camp Pendleton, San Diego County, California: in Gomes, Cari, ed., Waiting for Tsunami, Coastal Hazards of Northern San Diego County, California: San Diego Assoc. Geologists, Sunbelt Publications, p. 83–116.

San Luis Rey Indian Water Authority, n.d.(a), History. www.slriwa.org. Retrieved September 16, 2013, from http://www.slriwa.org/history

San Luis Rey Indian Water Authority, n.d.(b), Water Delivery. www.slriwa.org. Retrieved September 16, http://www.slriwa.org/delivery

San Luis Rey Indian Water Rights Settlement Act. 1988. Public Law No. 100-675, 100th Congress, 102 Stat 4000-4011. www.gpo.gov. Retrieved September 16, 2013, from http://www.gpo.gov/fdsys/pkg/STATUTE-102/pdf/STATUTE-102-Pg4000.pdf

Sierra Analytical Lab, 2012, Sample Site LCTS1: Report to Los Coyotes Band of Indians Environmental Office, 13 p.

Sparkman, P.S., 1908, The Culture of the Luiseño Indians: Univ. Calif. Pubs in Amer. Arch and Ethnology, v. 8, no. 4, Ramona, California, Ballena Press, 234 p.

Stephens, C.R., 2009, Age, composition and emplacement of the Santiago Peak Volcanics, Southern California (abs.): Geol. Soc. Amer. Abs. with Programs, v. 41, p. 644

Strong, W.D., 1929 [1974], Aboriginal Society in Southern California: Banning, CA, Malki Mus. Press, 258 p.

Sutton, M.Q., 2011, The Palomar tradition and its place in the prehistory of Southern California: Pacific Coast Arch. Soc., v. 44, p. 1–74.

Treganza, A.E., 1942, An archaeological reconnaissance of Northeastern Baja California and Southeastern California: Amer. Antiquity, v. 8, p. 152–163.

True, D.L., 1958, An early complex in San Diego County, California: American Antiquity, v. 23, p. 255–263.

True, D.L., 1966, Archaeological differentiation of Shoshonean and Yuman speaking groups in Southern California: Unpub. PhD Dissertation, Univ. Calif., Los Angeles.

True, D.L., Meighan, C.Q., and Drew, Harvey, 1974, Archaeological Investigations at Molpa, San Diego County, California: Berkeley, CA, University of California Press, 163 p.

U.S. Environmental Protection Agency. 1995. Manual, Ground-Water and Leachate Treatment System. EPA/625/R-94/005. 132 p. nepis.epa.gov. Retrieved September 15, 2013, from http://nepis.epa.gov/Adobe/PDF/30004NRU.pdf

Weston Solutions, Inc. (Weston), 2005, San Diego County Municipal Copermittees 2004-2005 Urban Runoff Monitoring. Final Report. Prepared for County of San Diego. December. www.projectcleanwater.org. Section 5, San Luis Rey River watershed management area, retrieved September 14, 2013, from http://www.projectcleanwater.org/pdf/science_mon/04-05monitoring/section_5_slr.pdf

Luiseño Indian structures on Palomar. (Robert Asher photo, from Brueggeman, 2008.)

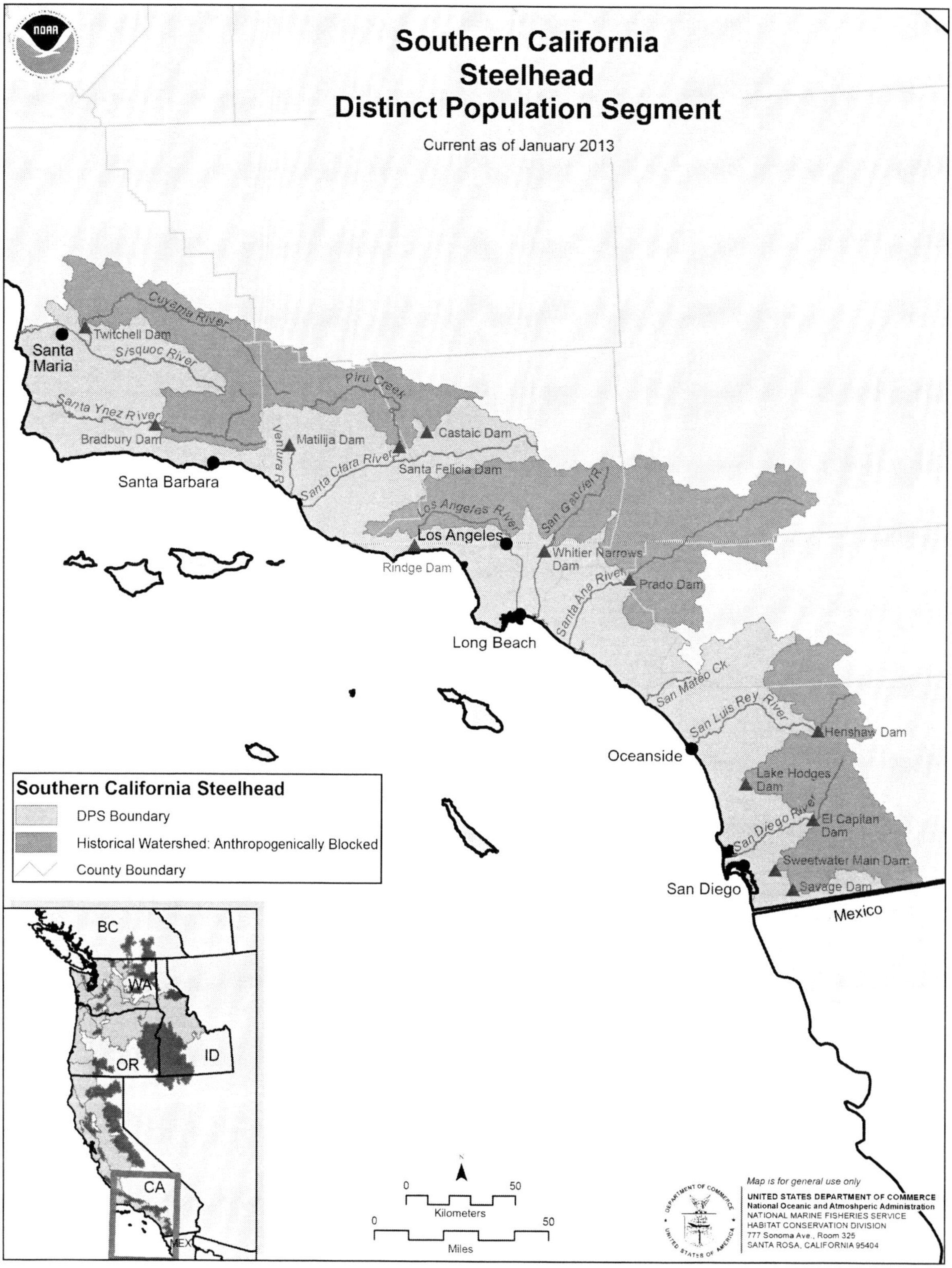
Southern California Steelhead Distinct Population Segment
Current as of January 2013
Cuyama River
Twitchell Dam
Santa Maria
Sisquoc River
Piru Creek
Santa Ynez River
Bradbury Dam
Ventura R
Matilija Dam
Castaic Dam
Santa Clara River
Santa Felicia Dam
Santa Barbara
Los Angeles River
San Gabriel R
Los Angeles
Whitier Narrows Dam
Rindge Dam
Santa Ana River
Prado Dam
Long Beach
San Mateo Ck
San Luis Rey River
Henshaw Dam
Oceanside
Lake Hodges Dam
San Diego River
El Capitan Dam
Sweetwater Main Dam
San Diego
Savage Dam
Mexico
Southern California Steelhead
DPS Boundary
Historical Watershed: Anthropogenically Blocked
County Boundary
BC
WA
ID
OR
CA
MEX
0
50
Kilometers
0
50
Miles
Map is for general use only
UNITED STATES DEPARTMENT OF COMMERCE
National Oceanic and Atmospheric Administration
NATIONAL MARINE FISHERIES SERVICE
HABITAT CONSERVATION DIVISION
777 Sonoma Ave., Room 325
SANTA ROSA, CALIFORNIA 95404

Rainbow Trout: From Kaibab Plateau to Kenai

Bruce Campbell bcampbell32@cox.net

What are Southern California's native rainbow trout and why are they worth preserving? I hope to answer these questions in the remarks that follow. Although I don't intend my remarks to be of scientific merit, the information I'm sharing comes from both scholarly sources and my personal observations and experiences. If you're interested, I provide references where I have borrowed from other authors.

EVOLUTION

Let's start with an overview of the evolutionary history of salmonids along the Pacific Coast of North America. *Oncorhynchus mykiss*, also known as rainbow trout, coastal rainbows, and redbands, began its journey about the same time the ancestral Colorado River began rapidly cutting its canyon through the western Kaibab Plateau, some 6 million years ago. Around the same time, a rift opened along the eastern edge of the Pacific plate, cutting off the Colorado from the sea and creating the pluvial Lake Chapala (Behnke 2002). Meanwhile, an early salmonid gave rise to two species: proto-rainbows and proto-cutthroats. The proto-cutthroats expanded their range into the Great Basin of western North America, and gave rise to more than a dozen subspecies now extant in the streams that once drained into Lake Bonneville. At about the same time, the proto-rainbows began expanding their habitat southward down the Colorado River into the streams feeding the pluvial lake that would become the Sea of Cortez, and probably used the Sea of Cortez as a place to feed and grow large, much as modern steelhead do. Some of these stay-behind rainbows still remain in streams that rise out of Sierra San Pedro y Martir, such as Rio Santo Domingo. As the rift expanded, eventually the Sea of Cortez opened to the Pacific and these highly migratory proto-rainbows began colonizing the west coast of North America.

The northward spread of these fish was halted by Pleistocene glaciations, and the early migrants became resident in the pluvial lakes of the northwest. They evolved into the redband strain of *O. mykiss,* found today from northern California (upper Sacramento River and tributaries) throughout central Oregon. Like present day cutthroats of the southwestern United States, they became primarily stream resident and largely lost the migratory habit. Those fish found in streams along the Oregon coast became today's coastal rainbows.

With the retreat of the Wisconsin glaciation, the proto-rainbows found south of the Wisconsin ice sheet again spread northward, becoming the coastal rainbows and steelhead found today from Oregon northward to Alaska's Kenai Peninsula.

Adaptations occurring following these geographic and geological changes served to isolate populations, ultimately resulting in speciation. Thus, modern fisheries biologists recognize at least three ecological zones for migratory *O. mykiss*: the Northern Pacific Steelhead Ecological Significant Unit (ESU), from the Columbia River and tributaries northward; the Central Coast ESU (from Carmel River to Oregon's Nehalem River), and finally the Southern Steelhead ESU, from northern Baja California to the streams feeding Morro Bay. There is continuing debate over whether the Southern Steelhead ESU should be further subdivided.

NATURAL HISTORY

Most of us are familiar with the life history of Pacific salmons. The Pacific salmons are characterized by being hatched and reared in clean, cold water streams, migrating downstream to the estuary, where smolting occurs, and then spending one or more years in the ocean feeding on herring, smelt and other prey species. At the end of their ocean-living period, adult Pacific salmons return to their natal streams to mate, spawn, and die. In many streams, especially those of British Columbia and Alaska, dead salmon form the base of the food chain that feeds invertebrates which then feed salmon fingerlings and smolts. (I know of no other life cycle where a species is at both the apex and the bottom of the food chain.)

We rarely think of rainbow trout as a salmon, yet it is: all salmon species are members of the genus *Oncorhyncus*. Rainbows differ in several primary physiological traits. First, there is always a part of an *O. mykiss* population that strays from its natal stream when spawning. The tendency to stray follows a geographic cline, decreasing from more than 50 percent in the south to less than 1 percent in the north. Second, temperature tolerance follows a bimodal distribution. In the northern part of *O. mykiss* range, 80 degrees Fahrenheit (°F) triggers flight responses – fish seek out cold springs or tributaries, or die. In contrast, southern California steelhead (*O. mykiss irideus*) has developed a remarkable level of heat tolerance. I personally have seen these fish actively feeding when the water temperature was 86°F, and I have seen them resting in water as warm as 90°F. Finally, *O. mykiss* in its northern ranges typically spawns in May to June; fish and wildlife departments in those areas have institutionalized fishing closures during this time to protect returning spawners. By contrast, here in southern California, our native trout enter streams (when and where possible) in January and February to spawn in March and April.

Anadromous (sea-run) rainbow trout can be broadly categorized into two groups, based on how long after migrating to the ocean the fish return to spawn for the first time. The first group is "one winter" steelhead, which returns for the first time after one year in the ocean. This group includes all steelhead south of the Carmel River. "Two winter" steelhead encompass all steelhead north of the Columbia River, and includes both "winter-run" and "summer-run" fish.

These two groups represent two broad adaptations to riparian conditions along the Pacific Coast. These conditions include drought, water temperature, watershed fire, and timing of peak flows. North of Carmel River, precipitation is more abundant and more uniformly distributed, fewer droughts, resulting in larger river, cooler river-forest systems that are naturally more resistant to catastrophic wildfires. Two winter fish predominate because fattening up for an extra year is needed to negotiate heavier stream flows, and conditions in the river are likely to be the same from one year to the next.

Frequent multiyear drought, uneven distribution of precipitation, high water temperature, and periodic watershed fires (anthropogenic or lightning caused) typify the riparian corridors draining the coasts of the Southern California Bight. These conditions have produced a salmonid that is well adapted to persisting in a hostile environment. Genetics may have a role in their persistence: morphologically, southern California steelhead progeny that become resident exhibit neoteny; they retain prominent parr marks (thumb-print shaped dark areas along the flank of the fish) and white tipped fins even as reproductive adults (Figure 1), and have two dark spots in the iris. Highly productive stream environments, characterized by alkaline pH, averaging approximately 8.3, with a large biomass and diverse benthic macroinvertebrate population (Isham et al. 2009), contribute to unusually rapid growth of young of the year, favoring a "one winter" fish.

Research by Girman and Garza (2006) shows a lack of genetic clustering in samples taken from geographically close watersheds, consistent with a high degree of gene flow, that is, straying and successful reproduction. This factor makes these fish better able to survive seasonal flashiness: if one watershed is damaged, adults in the ocean are likely to ascend neighboring streams where conditions may be more suitable for spawning. Unfortunately, it does nothing to protect them from anthropogenic destruction of reproductive habitat (e.g., dams, diversions, and groundwater removal). In another set of genetic analyses, Girman and Garza (2006) also found support for a recent reduction in genetic diversity consistent with widespread, dramatic decreases in effective population size. These findings are consistent with anthropogenic impacts occurring since the 1940s. At the same time, it found little variation in diversity between populations above dams and below dams, where *O. mykiss* is present in both locales.

SAN DIEGO DISTRIBUTION OF *O. MYKISS IRIDEUS*

Historically, anadromous *O.M. irideus* were found in all San Diego County's river systems, from San Mateo Creek in the north to Tijuana River in the south. San Mateo Creek hosted a run numbering in the thousands as recently as the 1950s, and supported a fishing camp until agricultural leases on Camp Pendleton began pumping ground water for irrigation. Escapes of exotic predatory fishes (principally largemouth bass and brown bullheads) from upstream impoundments severely impacted the survival of fingerlings in the deep pools of the upper drainage. In good water years, fish continue to ascend San Mateo Creek, with the most recent event occurring in 2002. Poor water conditions preventing the smolting of the 2002 young of the year, and ultimately led to their demise.

Early writers mention the presence of trout in the upper San Luis Rey River near what is now Warner Ranch, and unpublished interviews of Kumeyaay elders relate how the indigenous peoples used the fish as a food resource. Remnant resident populations persist on the slopes of Palomar Mountain in Pauma Creek as well as West Fork of San Luis Rey. These fish were able to survive the Poomacha Fire of 2007 (Figure 2), probably by sheltering in the deeper pools.

Downstream migrants from Pauma Creek

Figure 1. *Adult* O. mykiss *showing retention of parr marks (Houser 2012)*

Figure 2. *Pauma Creek canyon during Poomacha Fire (Greenwood 2007)*

continue to be observed in the vicinity of the Highway 78 crossing (Figure 3). In addition, large adults (approx. 18 inches) were observed in Aqua Tibia Creek in 2005, and in neighboring Gomez Creek in 2006.

California Department of Fish and Game, now Department of Fish and Wildlife (DFW), recorded probable steelhead catches in San Dieguito River (1946). Currently, all spawning and rearing habitat is blocked by Hodges and Sutherland dams.

San Diego River and the Sweetwater River probably formed a steelhead cluster because they shared San Diego Bay as a common estuary until historical times (San Diego River's old channel can still be seen separating Liberty Village and U.S. Marine Corps Recruit Depot). San Diego River also supported a steelhead population in the recent historical past. Like San Mateo Creek, Boulder Creek had a fish camp operation (near the Boulder Creek crossing of Eagle Creek Road) until the 1940s, when the construction of impassable

Figure 3. *Young of the year captured in Pauma Creek at Hwy 78 (Greenwood 2007)*

dams halted the migration of adult fish. Remnant populations remain in Boulder Creek, as well as nearby Cedar Creek.

Sweetwater River once held a large steelhead run as well, which was halted by construct of two dams along its main stem (Loveland Reservoir and Palo Verde Lake). Unlike the other San Diego streams, Loveland reservoir has never been stocked by DFW, but catches of large rainbow trout as recently as 2006 show that Loveland Reservoir is deep enough to remain cold throughout the year, and that trout survive in Loveland in sufficient numbers to repopulate headwaters of the stream, most recently after *O. mykiss* was wiped out from the headwaters by the Cedar Fire in 2003. Genetic analysis by Girman and Garza (2006) showed the fish had steelhead characteristics, and may have been part of a San Diego River "cluster." Figure 3 shows a rainbow trout caught and released in the upper Sweetwater in early 2012. Note the gravid condition of this female trout, as well as the faint parr marks, white fin borders, and twin eye spots. Naturalist observations (Greenwood and Campbell, 1995, 1998) have shown that rainbow trout in Sweetwater River spawned almost two months earlier than counterparts farther north (February –March vs. May-June). This is strong evidence for a genetic separation of the strains.

From the Tijuana River, there were reports of rainbow trout (not identified as steelhead) in small numbers as late as 1948. Pine Valley Creek, a Tijuana River tributary was stocked in 1995 and 1999, but these appeared to have died out. A naturalist made catches of large rainbow trout (17.5 inches in late May 1995; 18 inches in May 2005, and 16 inches in April 2009). These fish are absent from the upper portion of stream after June, and may represent a spawning run from the Pine Creek arm of Barrett Lake. Cottonwood Creek, another Tijuana River tributary, was stocked in 1995, 1999, and 2003; large fish were subsequently taken as recently as 2001 but have not persisted to the present. There is good habitat for trout from Morena Dam downstream to Hauser Creek, and large fish have been taken. There is no evidence as to the origin of these fish.

THE FUTURE

U.S. National Marine Fisheries Service (NMFS) successfully petitioned for listing the southern California steelhead (*O. mykiss nelsoni*) as an endangered species, which requires California's Department of Fish and Wildlife to bar all "take" of these fish. Anglers *must* release all fish caught unharmed, without removing the fish from the water. Unfortunately, NMFS's jurisdiction over anadromous fish ends at the base of each dam, and the U.S. Fish and Wildlife Service (FWS) has jurisdiction over all *O. mykiss* found above dams, and considers them to be resident breeding populations and not endangered. Moreover, FWS does not regard these resident fish as part of the *O. mykiss irideus* subspecies, even though the scientific evidence (Girman and Garza 2006) shows that the fish above and below the dam are genetically identical. As a result, there

Figure 4: *Gravid female* O.m. irideus *from Sweetwater River (Houser 2012)*

are no protections, other than "Wild Trout Stream" designations, for the largest part of the subspecies' gene pool. DFW isn't compelled to take action to preserve this part of the region's biological heritage.

California Supreme Court ruled against DFW in a judgment that restricts stocking of trout outside their "native" drainages. This ruling fails to consider that straying of migratory rainbow trout is most common in the Southern Steelhead, and decreases steadily in a northward cline. The narrow definition of "watershed" prevents restocking Sweetwater River progeny in the San Diego River, even though they are likely part of a San Diego River genetic cluster (Girman and Garza 2006).

In the face of likely global warming, *O. mykiss* faces serious threats to survival in the northern parts of its range. At least once in the species' past, *O. mykiss nelsoni* spread to marginal habitats, and its northern progeny no longer found survival value in tolerating high temperatures. If we fail to preserve the *O. mykiss irdeus* gene pool, we may be eliminating the best prospects for survival of the entire species.

REFERENCES

Behnke, R. J. 2002. *Trout and Salmon of North America*. New York: The Free Press.

Derek Girman, John Carlos Garza. 2006. Population structure and ancestry of *O. mykiss* populations in South-Central California based on genetic analysis of microsatellite data. NOAA Southwest Fisheries Science Center, 110 Shaffer Rd. Santa Cruz, CA 95060, November .

Greenwood, A. 1995. Oral history of trout fishing in San Diego County. Personal communication.

Greenwood, A. 2007. Poomacha fire photos. Personal communication.

Houser, C. 2013. Post Cedar Fire trout photos 2012. Personal communication.

Isham, B., Owen, D., Gerlinger, T., and others. 2009. *Bioassessment Monitoring of San Mateo Creek Waters*. Weston Solutions, Inc., Carlsbad, California.

Nate Harrison. Robert Asher photo.

Nate Harrison, George Doane, and the Roads to "Smith Mountain"

The following are selected chapters excerpted from **Palomar – From Tepee to Telescope**, *an excellent volume of "facts stories and pictures of the days when Palomar was different." Originally authored by Catherine M. Wood (Frye & Smith, 1937), it was revised by Peter Brueggeman (Premier Publishing, 2008), the former library director, now retired, at the Scripps Institution of Oceanography. These excerpts shed interesting light on life on Palomar (formerly Smith) Mountain at the turn of the 20th century.*

Considering we geologists tend to plot events on the scale of geologic time, in units of thousands to millions of years, these colorful characters are part of modern Palomar Mountain. It should be noted that these historical accounts do include some allusions to racial prejudices that were wrong then and are wrong today. However, to obscure them in any way would be the same as claiming such prejudices never existed. Please enjoy these excerpted selections on Nate Harrison, George Doane, and the Roads to "Smith Mountain." – Brian J. Olson, editor

THE FIRST WHITE MAN ON THE MOUNTAIN

Nathan Harrison, a negro, who always called himself "the first white man on the mountain" was a noted character, enjoyed by all who knew him. He was familiarly called Nate, or Uncle Nate, and because he lived so many years at a turn in the road on the west grade, that series of hairpin turns has been officially named "Nigger Grade." He was a fluent talker, but not about himself, unless encouraged by a little whiskey. As nearly as can be ascertained, he and his master came from Kentucky by ox team, stopping three or four months at Sedalia, Missouri, waiting to join a wagon train west. About the time they reached California, a movement was on foot in Los Angeles to break a road over the Tejon Pass to the north. Money was subscribed and bullocks furnished, and Nate drove an ox team with the first wagon train over the Pass. Cleland's "History of California" says: "In September, 1854, people of Los Angeles raised $6,000 for the construction of a wagon road between their city and Ft. Tejon." That seems to check with Nate's story. He and his master went up to the mining district near Merced, where his master died. Uncle Nate went down to the San Gabriel Mission for a time and later found his way to the foot of Palomar, where he took up a claim at Rincon.

He spent his summers in Doane valley, but did not take up a claim there, so the valley was not named for him. He lived with the Indians at times, and occasionally helped the early settlers with their sheep. Finally he sold his Rincon property and proved up on a claim on the west end of the mountain, past which the west grade was built about thirty years later. It was his custom to bring a bucket of water from his flowing spring for the horses when he heard a team laboring up the steep grade. He was usually rewarded with a bit of money or some food. Later the county dug a well and put in a pump at the roadside where Nate awaited the travelers. Now the well is dry and someone stole the pump, but it does not matter, for few horses travel the grade, and the modern automobile can make the trip without extra water.

Nate could neither read nor write, but he had a good memory for faces and events. He could not understand much about sums of money, it seems, for when a young Indian wanted to buy one of his horses, Nate asked $150 for it. The horse was not worth it, but it made no difference. The price was $150. The Indian said, "I wish Nate had never heard of $150. He wants $150 for every nag he has on the place."

Nate was "seventy-six years old next New Year's" for about twenty years. It is said that a couple of Palomar youths were responsible for his changing his age and birthday. They made him happy with whiskey, which he liked very much, told him he was a hundred and seven years old, and that his birthday was on the Fourth of July. And so it was thereafter! At that it may have been as correct as his previous recollection of his age. It is certain, however, that he was a grown man at the time of the Civil War.

Nate's cabin, about 1912. Photo by Averrett.

One prominent man of Palomar always gave Nate a bottle of whiskey at Christmas time, and the annual event was awaited eagerly by Nate. During the fall he would ask this man every time he saw him, "When is Christmas day?" One day after giving him his annual present, the donor happened to return to Nate's cabin. There sat Nate on the floor, one hand clutching the bottle by the neck, and he explained, "I'se a'goin' to sit and sip this here whiskey till it's a-l-l gone." The statement was not doubted.

Someone else remarked that he would like to put old Nate in a tank of water to see how many pairs of overalls he could soak off.

One of the men at a road construction camp near Nate's, picked a paper one evening and pretended to read a tale of how a fleeing murderer was headed towards the west end of the mountain

Nathan Harrison and his monument. Photo Mrs. J.B. Reece.

until he had Nate sitting around with a shot gun in his hand. The practical joker little realized that had a stranger appeared at that time, Nate surely would have taken a shot at him.

Nearly everyone who knew Nate remembers one or more interesting incidents about him. One was amused because Nate always pronounced the name of the town Escondido as "Skundido." To a San Diegan trying out his car on the steep grade in the days when motoring was an adventure, Nate said, "Awful hot mornin', Mr. Jackson. Keeps you in the collar all the way up the hill, don't it?"

Though people who knew him joked with Nate and often teased him, they were good to him. One family sent him fresh meat whenever they butchered. He seemed to enjoy the attention people gave him. One writer said, "Nate was a good-for-nothing, lovable old darkey, loved in spite of, not because of, certain qualities he possessed."

The story was always current that Nate "had money," but while it added to people's interest in him it was apparently unfounded. He lived alone in his squalid little hut, which his scrawny chickens sometimes shared with him until some kind hearted people thought he was no longer able to care for himself. He was invited for a ride by his well-meaning friends, which ended at the County Hospital in San Diego; he was not given a chance to take his pipe or any of his other little personal belongings, and always wanted to return to Palomar. It has been said that when the nurses peeled off the accumulated layers of overalls it was the beginning of the end, and when they bathed him, that was the end. Friends who visited him at the hospital, said he seemed happy and told them the nurses were good to him. He died there in 1920.

Largely through the efforts of Theodore Bailey of Palomar, funds were subscribed for a monument of native granite which stands at the turn in the road where Nate usually greeted travelers. The inscription on the plaque which was made at Jessop's Jewelry Store in San Diego, reads:

> Nathan Harrison's Spring
> Brought here a slave about 1848
> Died Oct. 10, 1920
> Aged 101 years.
> "A man's a man for a' that."

THE ROMANCE OF DOANE VALLEY

George E. Doane was the truly romantic character of Palomar Mountain, and seemingly enjoyed the role. He once owned the property known as Upper and Lower Doane Valley, which is now included in the State Park. Aside from his romantic attitudes, his chief distinction was a flowing beard, which covered most of his face and reached to his waist. He was a well-educated man from the San Francisco Bay region and was a hotel clerk in San Diego in early days before he went to Palomar. He was once voted the handsomest man in San Diego, so it is said, in some sort of contest, and his flowing moustaches were a point in his favor.

George E. Doane was one of the earliest white settlers on Palomar; his mother Nancy Doane, a frail little lady, had a claim nearby. She lived in a little cabin adjoining her son's, but each cabin was on a different parcel of land, with a covered passageway connecting the two. Nancy Doane would cook good meals for her son and his helpers, but did not eat with them. Nancy Doane died on the mountain; Doane bore her body in a buckboard down the steep mountain "slide" not far from where the present south grade now runs, and she was buried in Valley Center.

George E. Doane

Two brothers of George Doane were also on the mountain for a time [PB: one brother, John, figures in Valley Center history]. George's younger brother was a deaf mute, so it is said. One time he was returning from a trip below after supplies;

George Doane. This picture was taken shortly before his marriage. In the box on the wall were twenty-nine answers to his advertisements for a wife. Robert Asher photo.

about the time he reached the Mendenhall home in Malava Valley, night was falling and a snowstorm had started. The Mendenhalls tried to get him to stay all night, but, determined to proceed, he started on, over the rough, narrow road. Before he had gone far, the wagon overturned in a snow-covered ditch. He unhitched the horses and drove them on home, and in the morning George Doane came on horseback and rescued the supplies, but the wagon had to remain there the rest of the winter.

It was a standing joke throughout the county that Doane liked school teachers, and did not care who knew it. He was quite a poet, and amused his friends with this rhyme:

"Though I like doughnuts and clams, Still better I like the school ma'ams."

Someone told him he had better cut his beard if he expected to win one, — he replied that if she took him she would have to take his beard too! He liked to visit schools, and he used to go to Teachers' Institutes in San Diego, where on one occasion at least he came to the hotel dinner table attired in a dress suit. He often wore a small derby hat even when on Palomar.

He was supposed to have proposed to every unmarried woman who came to the mountain, all to no avail, apparently, for after living on Palomar over twenty years, he advertised for a wife. He received a number of letters and pictures, which

George Doane's cabin. Robert Asher photo.

were talked over with his mountain friends. On one occasion he asked a friend what he thought of "that bunch of heifers," referring to a group of pictures. The friend told him he thought some of them had been on the range too long.

Doane finally selected two as promising, a mother and her daughter, both open to marriage. By that time he had paid for Upper Doane Valley, which he had bought from Kitching, so, one year after he sold his cattle, he went to Louisiana to look over the situation. He married the daughter, about sixteen years of age, and brought her and a young negro maid back to the crude log cabin on Palomar. Later Mrs. Doane's mother and sister came west and homesteaded on what is known as the Hayes' place.

The romance caused a thrill of interest throughout the whole country. As Doane, his wife, and the negress went up the mountain, they stopped at Nigger Nate's a minute, and Doane, so the story goes, told Nate he had brought him a wife. Nate's reply was, "Which one?"

The negress, about sixteen years old and very large, wore no shoes and was designated by some of the mountain people as "Amy Nigger with the big feet," but Nate named her "Cubby" because she had feet as big as a bear's. Amy worked hard. She even pitched meadow hay with Doane, one on each side of the wagon. Being young and strong she could work faster than he could, and when she finished on her side, she would go around back of the wagon, lean on her fork and "cuss him out" because he could not keep up with her. Some of the boys of the vicinity used to go over to Doane Valley during haying time to enjoy the fun.

One day the Doane group came to Bailey's store from which they bought groceries, complaining of the quality of the food they had purchased the day before, but they were assured that the groceries were all right when they left the store. They went on home and learned later that Amy had tried to poison the family by putting lye in the lard, because they made her work so hard.

Mrs. Doane had her seventeenth birthday on the mountain. One day Doane came to the store with his flowing beard very black, and his shirt front also, where the beard had rubbed against it. His young wife, in an attempt to rejuvenate her elderly husband had dyed his grey beard with shoe blacking!

Mr. and Mrs. George E. Doane sold their property in 1905 and moved away. It is said that there was a divorce later. Amy was sent back to her home in Louisiana when the family left the mountain. Accounts differ as to what became of George E. Doane, but it is understood that he eventually left California.

PALOMAR WAS SMITH MOUNTAIN TO EARLY SETTLERS

Until the turn of the century, the accepted name for Palomar was Smith Mountain. That is the name which occurs on all county maps of previous date. The name was applied especially to the east end, where, on a spring-watered slope with a magnificent view over what is now Dyche Valley in the foreground, Warner's Ranch below, and desert mountains in the distance, Joseph Smith of San Diego settled. He was a very tall man, so was commonly called "Long Joe" or "Largo" Smith.

Old records indicate that he was in San Diego as early as 1853. He served on the County Board of Supervisors in 1856, 1857, 1865, and 1866. It may be assumed that Smith was a carpenter by profession, or at least proficient with his tools, for he at one time received ten dollars for making a coffin, but the price of coffins evidently went up, for a year or so later he was paid twelve dollars each for two coffins, one for an Indian and one for a Cholo (half-breed). Indians at that time were a problem throughout the county. There were hundreds of them, and unscrupulous people persisted in selling them whiskey, illegally, so coffins were often needed. He was also paid sixty-six dollars for benches for the Court House, then, of course, at Old Town.

In 1856 Joseph Smith was one of a committee appointed to consider a route to Vallecitos, the first Overland Mail Route, and the ensuing year was "Superintendent of the work on the new road to the desert." He was said to have been a little "hard-headed" in Board meetings sometimes, but "we vote for Smith because he is 'sound' on the road question." [Minutes of the Board of Supervisors, San Diego County-Book 1.]

In 1859 Smith and Ephraim W. Morse, later known as "San Diego's Ideal Citizen" and who was running a general merchandise store in Old Town at the time, located a cattle and sheep ranch on Palomar Mountain, where, according to the San Diego paper "they will not only raise grain enough for their own use but have plenty for sale." [*San*

The home of Joseph Smith. Mary Mendenhall Knox photo.

Diego Herald, April 2, 1859] Another note in the same paper mentions that hay and grain were getting scarce on the Overland Mail Route between Fort Yuma and El Paso. In 1861 Morse returned to San Diego to resume his former occupation of storekeeper, probably when the opening of the Civil War stopped the Butterfield Stages, but Joseph Smith remained on the mountain.

That Smith was a Palomar booster is indicated by the following incident:

"A good story is out on our friend Largo Smith, which is worth preserving, in a climatic point of view. His ranch is situated upon a mesa about 3000 feet high. Some of his friends were joking him on selecting such a spot, and, thought it must be very cold up there. 'Why Smith,' said one 'you must have perpetual snow on your place!'

"'Well, no' replied he, 'we have some snow, to be sure; but then, you see, it is not like the snow we have in the States; ours is a warm snow!'" [San Diego Herald, April 9, 1859]

Smith was appointed Road Overseer in the Agua Caliente township in 1861-1862, and Judge of the Plains in 1862 and 1863. This office in the San Diego district dates back as far as 1835. In that year Juan Bandini and others prayed for appointment of Judges of the Plains because of "repeated and scandalous robberies of cattle." [Index to Spanish Archives sent to San Francisco-Office of County Recorder, San Diego, Calif.]

The duties of the office as defined in 1857 by the County Board of Supervisors throw a bright light on economic and social conditions of the time. There were three Judges of the Plains for the county at large and one or more for each township.

Some of the duties were as follows:

"...to make and attend once a year a gathering of the Horses together and give notice of the time and place."

"It shall be the duty of all persons killing a beef to advise a Judge of the Plains, not a relative, before so doing, to keep the hide at least two days, so that any of the Judges of the Plains, or one authorized by anyone of them, may examine the same, and it is hereby made their duty to make such examination and keep a record of the same open at all times to the inspection of any citizen of the county under penalty of a $10 fine for the first offense or one month's imprisonment or both fine and imprisonment at the discretion of the Court...

"It shall be the duty of the Judges of the Plains to examine all calf pens and see that calves and cows correspond.

"It shall be the duty of each owner of a ranch to keep a copy of these rules posted upon his premises."

One hundred copies of these rules were ordered printed.

Joseph Smith built a four-room adobe house on his Palomar ranch, a barn, and smoke house, and raised sheep, hogs and cattle. He supplied bacon, ham and other commodities to the stage lines and to San Diego, using Indian labor, and ox-carts for hauling supplies up and down the steep east slope of the mountain over which he had broken a road.

On one of his trips below, Smith brought back to Palomar a man supposedly a deserter from a British ship, to act as foreman for the Indian laborers. This man murdered his benefactor, legend says, because of his infatuation for Smith's Indian housekeeper, but here is the account of the affair which was sent by the San Diego correspondent to the San Francisco Bulletin probably in the spring of 1868, for the Smith estate was administered by Ephraim W. Morse in the fall of that year.

"The citizens here were very much excited last week by learning that Joseph Smith, known everywhere as 'Long Joe Smith' has been assassinated at his ranch 'Palomar.' It seems an Englishman whose name I did not learn, for some reason unknown, shot Smith dead while working at a bench. The Englishman then started for another place to bring some wine Smith had contracted for, leaving the latter as he fell. On his return he found a number of neighbors at the house and being asked what had become of Joe, he gave no satisfactory answer, but finally, on being accused of murdering Smith whose body had been found before the murderer had returned, he confessed the deed. Quite a sum of money was found in the house untouched, so it is not known whether poor Joe was killed for money or not. Next morning the murderer was found hung to a tree in the neighborhood." [In the San Diego Historical Society Museum.]

Site of Joseph Smith's home, in 1928.

Old stage station, Warner's Ranch, in the vicinity of which Smith's murderer was hanged.

There was no San Diego paper to record the event, as The Herald was no longer in existence, and The Union began later that year, but rumor has it that in order to get the man to confess the murder it was necessary to let him feel the noose tightening around his neck several times, and that he was taken to Warner's Ranch, chained to a wagon wheel all night and hanged the next morning with about twenty- five men on the other end of the rope, one of them being Nate Harrison. A variation of the tale is that, as they stopped at Warner's, someone called the attention of the constable inside the store, and when he looked out to see how his prisoner was getting along, the man was hanging from a tree. Nearly every tree in the neighborhood of the old store has had its turn in being pointed out as the one on which the murderer was hanged. One story said it was customary to cut down a tree to which a person had been hanged and that it was done in this case.

Articles have been printed claiming that the name Smith Mountain came from the fact that Thomas Long "Peg-leg" Smith of Lost Gold Mine fame, once stopped at the foot of the mountain to secure water from a spring there, and later used the mountain as a starting point in his unsuccessful attempt to re-discover the fabulously rich mine. That would have made the name date back to 1827, but there is no available record of the application of the name Smith Mountain until the time of Joseph Smith, for whom old settlers agree the mountain was named.

The Division of Geographic Names, Washington, D.C., says: "In response to a petition from local citizens for a change in name from Smith Mountain, the name Palomar was officially adopted by this organization on December 4, 1901. Palomar was the original Spanish name for the mountain, meaning 'dovecot,' and occurred on early maps. The petition of the local residents stated that Smith Mountain was so designated 'some twenty years ago' by residents in the vicinity, because a man named Smith met a violent death there." (George C. Martin, Executive Secretary.)

Soon after the selection of Palomar as the site for the large telescope, possibly in an attempt to move the mountain closer to San Diego, a movement was started to change the name of Palomar to San Diego Mountain. To this the Institute of Technology objected, saying that the name Palomar appears on the Government Topographical maps and to change it would result in confusion. "For these and other important reasons we sincerely hope that Palomar Mountain will always retain its present beautiful name." (J. A. Anderson, Executive Officer.) That was all that was necessary. Nothing further was heard about changing the name.

Many years ago, when the change of name was being agitated, the following poem by a Young appeared in an Oceanside paper. Young is said to have enjoyed yearly camping trips to Palomar.

A PLEA FOR PALOMAR

Fell my oak and fell my pine tree, send my cedar to the mill,
Strip the tangled vine from off me, roll my boulders down the hill,
Grade my summit, fill my valleys, tear away my woodland pride,
Parcel me in City lots and run a railroad up my side.
Rule my streets with dull precision, block by block in order true,
Here a church and there a steeple, where the tiger lilies grew.
Mar God's handiwork about me, let my beauty be a myth,
Then, defaced and desecrated, call me after Mr. Smith!

But while yet the stately cedar sentinels the sylvan lawn,
While at times from yonder thicket peeps the nimble-footed fawn;
While the glory of the morning breaks on precipice and peak
And the winter sees my waters leaping down to Pauma Creek;
While the valley smiles beneath me, stretching westward to the main

Mile on mile of rolling pasture, green alfalfa, golden grain;
While I look on Catalina far beyond the ocean shore
And the gleam of sunny waters on the lake of Elsinore,
While I dominate the lowland, hill, and valley, near and far,
In my majesty and beauty, let my name be Palomar!

MOUNTAIN ROADS

The first road on Palomar was over the east end, the one broken through by Joseph Smith. It was noted for its steep "slides" of something like 30 per cent grade. Heavy flat-bottomed shoes of iron were tied to wheels of the clumsy ox carts then in use, when descending these steep pitches. The shoes acted as brakes by causing the wheels to slide instead of turn. Later a county road was constructed on the south slope near the route of the new grade. When no longer used as a road, the south slide became the "mail trail" up which the mail was brought several times a week on horseback, but the carrier walked up the steep parts.

About 1900, as better transportation facilities were demanded by an increasing population, the nine-mile Nigger Grade [now known as Nate Harrison Grade] was built over the treeless west shoulder. It was considered at the time a good road, but is now noted for its steep pitches and hairpin turns. A traveler recently returned from a European trip remarked as she gazed down on the series of switchbacks yet to come, "St. Gothard's Pass has nothing on this except the snow." Nate Harrison Grade offered a favorite climb in early

Nigger Grade, now known as Nate Harrison Grade. Robert Asher photo.

days of motoring for testing speed and endurance of various makes of cars, and became widely known in automobile circles.

It was formerly the custom when descending the mountain to tie a tree to the back of the wagon to act as a brake. The first cars also followed this practice, but it was stopped by the Forest Service. Discarded trees at the foot of Nate Harrison Grade supplied firewood for the Indians of the region for many years.

Trees tied behind cars served as brakes, 1912. Mrs. J. B. Reece.

Senator Ed Fletcher of San Diego, in his bicycling days in the early 1890s, pushed his bicycle to the top of the mountain over the south slide, and rode down on an apple wagon with a tree tied behind it. In 1908 he ascended the original east grade in his two-cylinder Maxwell, which had to back up the steep pitches in order that the engine might receive enough gas. If the exhausted car had to use a little mule power in addition to its own horsepower in order to make the entire grade, what of it? Fletcher has been working for a good road up the east end ever since.

Ed Fletcher in his two-cylinder Maxwell, 1908. Ed Fletcher photo.

The east grade, about fifteen miles long, not so steep as Nate Harrison Grade, was not fully improved until about 1925. A few years before that, one resourceful young woman coaxed a gasping Ford over a renowned steep pitch with a sharp turn at the end by blowing in the gas tank to increase the pressure while the driver worked feverishly with feet and hands to pilot the car.

The new south grade was built especially for the telescope's trip up the mountain. Downward from the crest, drooling, loose-lipped steam shovels gouged and spewed, eating out their own road as they carelessly thrust aside trees and boulders. Upward from the base of the mountain, men of the County Prison camp, paying their debt to society for minor offenses, worked their way to freedom. Because of its south exposure, this "Highway to the Stars" will afford access to the mountain when the other two roads are snow-blocked on their north slopes.

A NARROW ESCAPE

Henry Edwards Huntington, the railroad magnate, about 1905, planned extensive developments in Southern California, including an electric line between Los Angeles and San Diego, with water projects in the Palomar region. To investigate the possibilities of the area, he and his party took a trip up Nate Harrison Grade, across the top of the mountain and down the east slide. Their conveyance was a three-seated rig drawn by four horses; their driver was Louis Salmons.

To ascend Nate Harrison Grade as it used to be, behind four lively horses, must have been exciting to say the least, but all went well until the party reached the east "slide" and there the brakes gave way. The driver managed to hold back the horses until the members of the party could scramble out and grab the wheels. These they managed to tie in such a way that the remainder of the descent was made in safety.

Meals for the three days' outing had been put up separately by H. Jevne Company of Los Angeles, so at lunch time, following the eventful trip, Huntington built a fire, made coffee, and all enjoyed an excellent picnic lunch.

Water rights at the site of Lake Henshaw were secured and Doane Valley purchased, but the electric line was never built, as there were difficulties over franchises. It has been claimed that Harriman interests blocked developments. After Huntington died, Charles Adelbert Canfield [PB: Edward Doheny's first and most successful partner in mining and oil development] took the Doane Valley property and was planning extensive improvements, which were cut short by his death in 1913. His daughter, Daisy Canfield Danziger, married to Antonio Moreno, a well-known Hollywood actor, inherited the property and held it until it was purchased for the State Park. Had the developments gone as planned, would the Telescope have been lost to Palomar?

Former Ocean View Camp Store, put in at the "junction" prior to the subdivision of Birch Hill into camp sites (circa 1920). This building was replaced by a new one in 1936.

IV. APPENDICES

Archaeological Investigations at Molpa

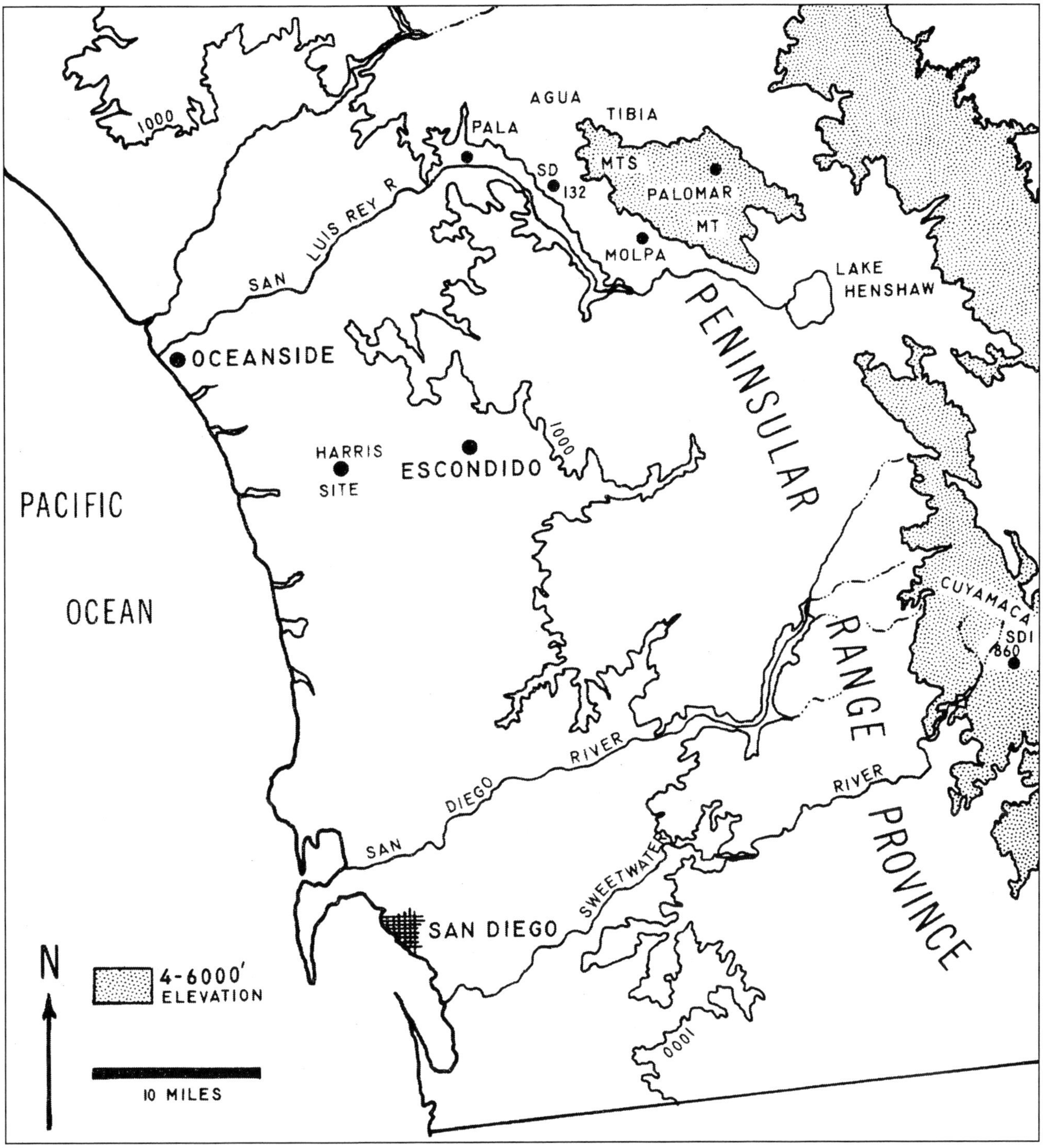

Map 1. *Western San Diego County and general area discussed in Molpa report.* *See road log, page 18.*

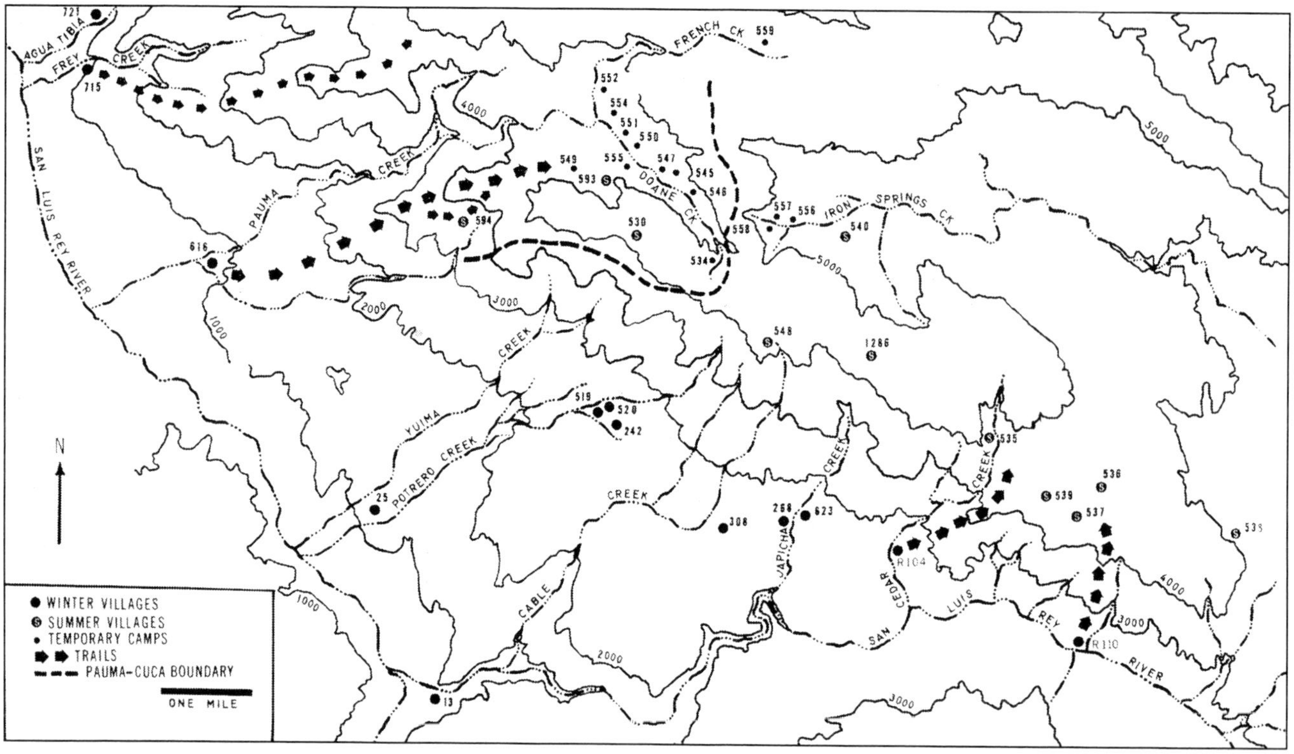

Map 2. *Indicates location of major sites in middle San Luis Rey River drainage. This includes both summer and winter villages, and the trails connecting them when these are known. Source (True et al, 1974). Per True's bi-polar fission or fusion model of habitation, groups are believed to have congregated during the winter months at main villages in the river valley, such as Molpa (SDI #308), and then dispersed during the summer to higher elevation sites on Palomar Mountain. Numbered sites correspond to the Site Trinomial (San Diego County or SDI) number recorded by the California Historical Resources Information System (CHRIS) and the South Coastal Information Center (SCIC).*

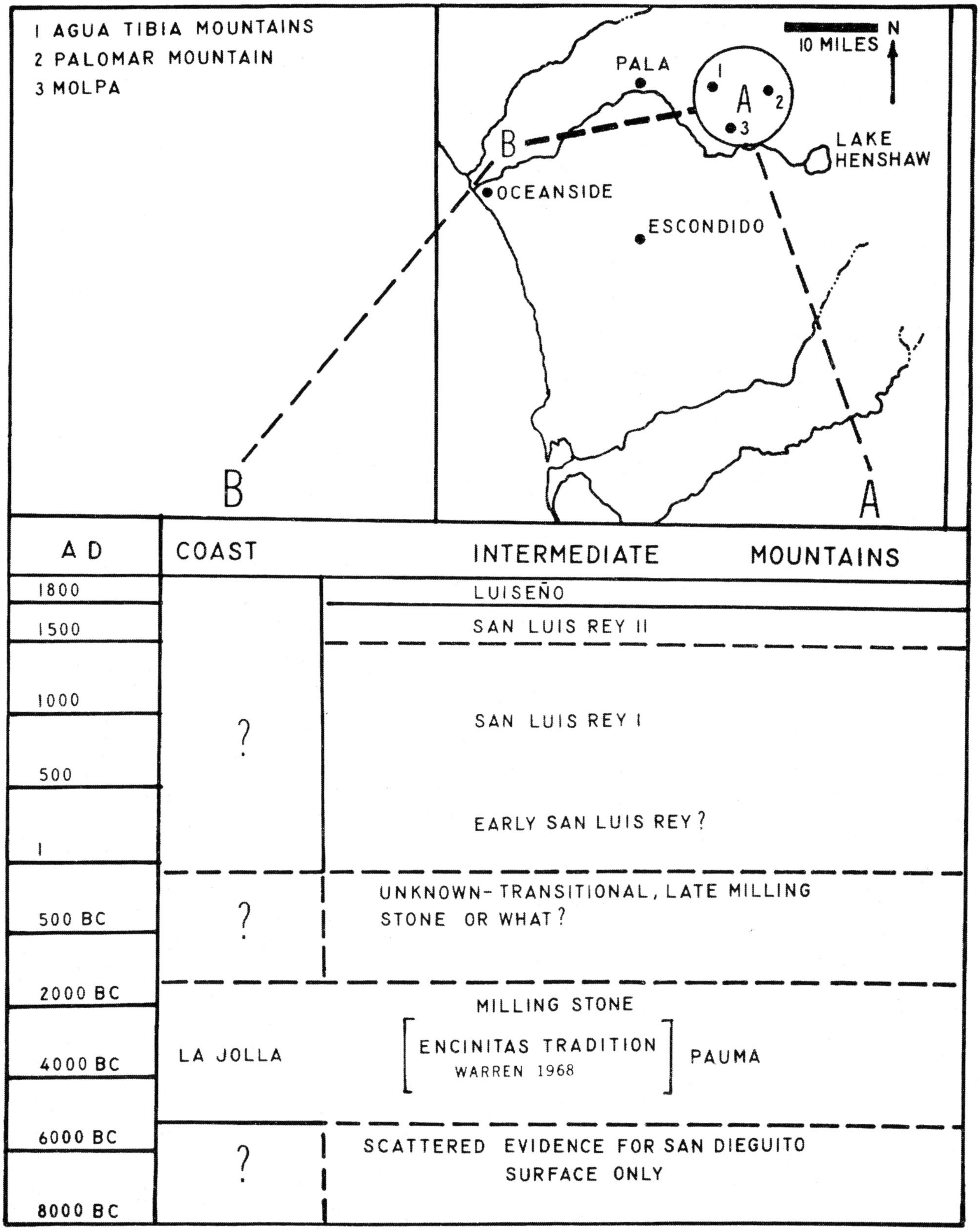

Figure 1. *Suggested time and cultural relationships in San Diego County, California, along a transect extending from the Palomar Mountain region to the coast near the city of Oceanside. This is all within the territory designated as San Luis Rey.*

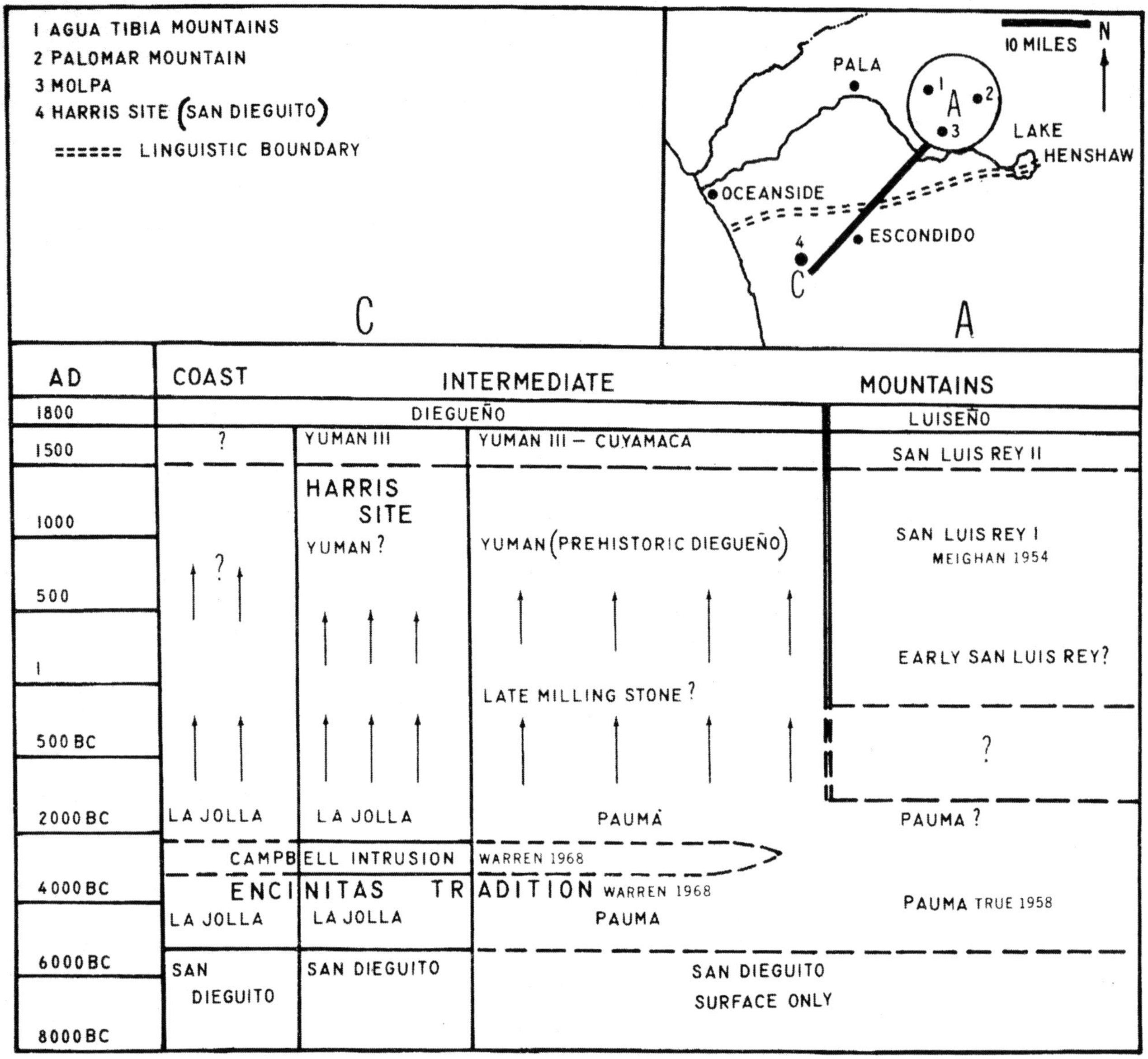

Figure 2. *Suggested time and cultural relationships in San Diego County, California, along a transect extending from the Palomar Mountain region to the coastal plain south and west of the city of Escondido. This transect includes territory belonging to both the Luiseño (San Luis Rey) and the Diegueño (Cuyamaca) peoples.*

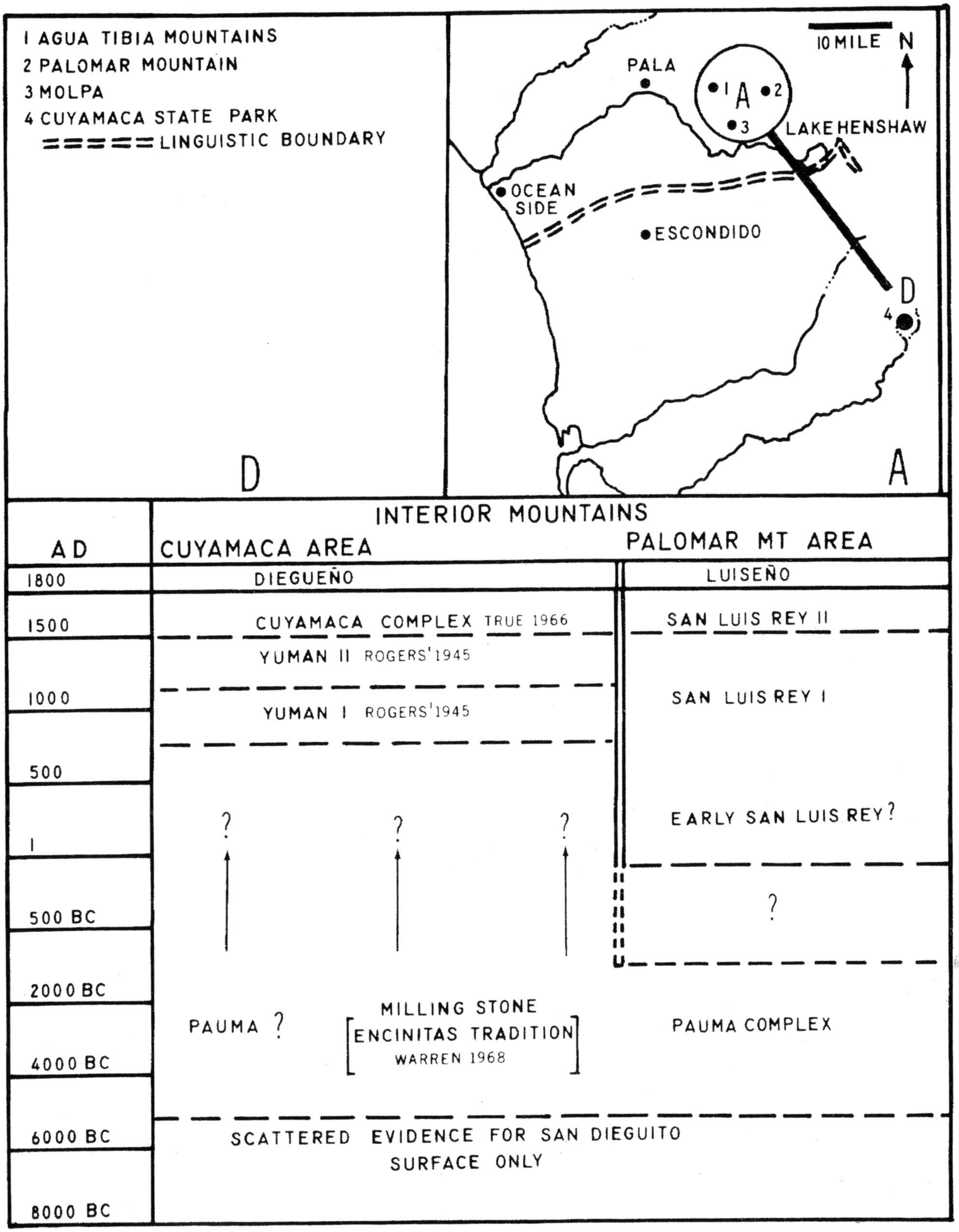

Figure 3. *Suggested time and cultural relationships in San Diego County, California, along a transect extending from the Palomar Mountain region to the Laguna Mountain region in the vicinity of Cuyamaca Rancho State Park. This transect includes territory of the San Luis Rey and Cuyamaca complexes.*

Distribution of Artifacts by Material

	Xtaline quartz	Basalt	Chert	Chalcedony	Felsite	Obsidian	Other	Total
Projectile Points								
Type 1	143	13	5	2	4	8	10	185
Type 2	61	1	-	-	-	1	1	64
Type 3	23	1	1	-	-	1	-	26
Type 4	-	-	1	-	-	-	-	1
Type 5	5	2	-	-	-	1	1	9
Type 12	3	1	-	-	-	-	1	5
nondiagnostic	109	7	2	-	4	2	9	133
Knives								
Type 1	12		1	-	-	-	1	15
Type 2	7	2	1	-	-	-	2	12
Type 3	4	-	-	-	1	-	-	5
Type 4	9	-	-	-	1	-	1	11
Type 6	3	-	-	-	-	-	-	3
irregular flake	-	2	-	-	1	-	-	3
utilized flake	2	-	-	-	-	2	-	4
nondiagnostic conventional forms	22	2	2	-	1	1	1	29
Drills	11	-	-	1	-	-	-	12
Scrapers								
domed	-	2	-	-	1	-	2	5
keeled	-	1	-	-	-	-	-	1
flake	-	2	-	-	-	-	3	5
cortex-based	-	2	-	-	-	-	-	2
nondiagnostic	-	2	-	-	-	-	-	2
Worked flakes	50	1	1	-	1	6	-	59
Scraper plane	-	1	-	-	1	-	-	2
Chopper	-	1	-	-	-	-	-	1

Total number of chipped stone artifacts: 596
Chipped stone artifacts made of crystaline quartz: 464
Percentage of chipped stone made on quartz: 78%

Figure 4. *The Molpa Archaeological Site (SDI-308) is an important "type site" that helps define the archaeological assemblage in the San Luis Rey River drainage. Excavations of the site were performed between 1955 through 1957 by field classes in archaeology from the University of California, Los Angeles, under the direction of C.W. Meighan. Over eight hundred artifacts (not including potsherds) were recovered. Of these, 596 were some sort of chipped or shaped stone implement, defined in the table above based on geologic material. The crystalline (aka Xtaline) quartz is clearly the most common composition of worked flake implement. Presumably, pegmatites provided quartz workable into sharp, durable cutting tools. Other compositions, such as basalt and obsidian, are indicative of import. Note that the "types" of projectile points and knives were only differentiated based on subtle variations in size and shape. Source (Table 4, True et al, 1974).*